Site cost control in the construction industry

Site cost control in the construction industry

J. Gobourne MCIOB, MMS
Development Engineer,
Sir Alfred McAlpine and Son (Southern) Ltd

Butterworth Scientific
London Boston Durban Singapore Sydney Toronto Wellington

All rights reserved. No part of this publication may be reproduced
or transmitted in any form or by any means, including
photocopying and recording, without the written permission of
the copyright holder, applications for which should be addressed
to the Publishers. Such written permission must also be obtained
before any part of this publication is stored in a retrieval system of
any nature.

This book is sold subject to the Standard Conditions of Sale of
Net Books and may not be resold in the UK below the net price
given by the Publishers in their current price list.

First published, 1982

© J. Gobourne, 1982

British Library Cataloguing in Publication Data

Gobourne, J.
 Site cost control in the construction industry.
 1. Building sites
 I. Title
 690'.068'1   HD9715

ISBN 0 408 01222 6
ISBN 0 408 01122 X Pbk

Photoset by Butterworths Litho Preparation Department
Printed in England by Butler & Tanner Ltd, Frome, Somerset

Preface to second edition

In the first edition of this book I looked into the various techniques of monitoring costs on construction sites. This now forms Part 1 of the second edition whilst Part 2 looks at the techniques of effecting a cost improvement.

It is a sign of the times that since the writing of the first edition only 10 years ago, labour costs have increased by about 500%. Competition to win contracts is undoubtedly more fierce, profit margins have been slashed, bankruptcies and takeovers are commonplace. In short, the industry has never had it so bad! Nor will the eventual upturn ease our burden; with a diminishing pool of skilled labour, companies will be caught in the cyclic trap of having tendered for contracts on fiercely competitive margins and then have to carry out the work when labour is at a premium.

Nothing must be left to chance. Resources must be rationed, wastage stamped out, gang sizes balanced, labour motivated, plant fully utilised, decisions must be thought through beforehand and then monitored during construction. Few will deny that there is scope for improvement on any building site but the techniques of bringing about the improvements take willpower, energy, retraining and of course increase the dreaded paperwork and the all important overheads. Cost control must be cost effective to be worthwhile but we can all, I am sure, bring to mind the more obvious situations on site where with a little more 'control' company money would not have been poured down the proverbial drain.

Such situations themselves are proof enough that cost control is necessary; but further than that there are the numerous unseen losses which only become apparent when the detailed techniques recommended in this book are followed. Only then can we claim to be getting to grips with productivity and the control of costs on a construction site.

J. Gobourne
1982

Publishers' note

Names of personnel given in the various task sheets, time allocation sheets, etc., throughout this book, are completely imaginary and bear no relationship to actual persons living or dead.

Acknowledgement

Acknowledgement is due to the Construction Industry Consul of the Universal Esperanto Association in Stockholm, Inĝeniero Per Törnegren Fakdelegito de UEA por Konstrutekniko, who provided such useful information, both in Esperanto and in English, on the piece rate system and work study data bank of the Federation of Swedish Building Employers.

Preface to first edition

In this book I have described the principles of cost control as applicable to the construction industry, and have shown how these principles can be applied to measuring and controlling the utilisation of labour, plant materials and overheads on a construction site.

No two companies are likely to evolve absolutely identical costing systems, because they are constantly governed by factors outside the costing system itself, such as accountancy procedures, payroll systems, methods of estimating, degree of planning, incentive schemes, use of computer, staff availability, company policies and types and density of construction undertaken. For this reason the book continually refers to variations of the point under discussion in order that the reader, upon completion, can tailor-make a cost control system to suit his own company's conditions and to incorporate whatever additional factors may be required by that company, e.g. incentive bonus payments, feed-back of output data to estimators, profitability of various trades, ratios of tradesmen to labourers, exclusion of overheads and inclusions of staff.

The examples of cost control systems given in the book, although primarily intended for Institute of Building Associate examination students, can be seen to apply equally to many aspects of site construction work, such as civil, mechanical and electrical, the variation being one of technical phraseology rather than basic principles.

J. Gobourne

Wolverhampton, 1972

Contents

Preface to second edition
Preface to first edition

Part 1 Monitoring costs — 1

1. Introduction — 3
2. Contract accounts and prime costs — 7
3. Materials — 13
4. Sub-contractors — 18
5. Trade costs — 20
6. Unit costing — 41
7. Standard costing — 50
8. Monitoring expenditure — 57
9. Standards for measurable work — 68
10. Standards for site overheads — 80
11. Standard cost example — 86
12. Standard cost exercise — 114
13. Costing tips — 120
14. Costing of jobbing works — 133

Part 2 Cost improvement — 139

15. Introduction to cost improvement — 141
16. Action — 145
17. Work study – work measurement — 150
18. Work study – method study — 170
19. TYMLOG recorder — 181
20. Bulk earthmoving – scrapers or dump trucks — 195
21. Ready-mixed or site-mixed concrete — 203
22. Purchase or site cut and bend reinforcement — 209
23. Hours of work — 211
24. National increases — 227
25. Incentive schemes — 230
26. Job cards — 242
27. Weekly planning — 246
28. Pre-costing — 250
29. Computers — 254
30. Computer costing — 260

Index — 277

Part 1
Monitoring costs

Motto

Each minute wasted of a man's time is equal in cost to throwing away a house brick.

If all the minutes lost on a construction site were stacked up as bricks it would be easier to find them. It is knowing where to look that is the first hurdle in the control of costs.

Chapter 1
Introduction

Firm's accounts

A firm's annual profit and loss account provides the master control of all that firm's activities. If an unsatisfactory balance is shown year after year, it is plain that action is required to prevent eventual bankruptcy. One would think that the problem would never be allowed to go that far; yet the building and civil engineering industry is noted for its high rate of bankruptcies, not only among the smaller builders but also among the larger companies.

Contract accounts

By separating accounts for individual contracts, both of money earned and money spent, an account can be built up of the profitability of each individual contract. Costs not applicable to any one particular contract, such as head office overheads, directors' fees, bank charges, taxation, etc., can be apportioned to each contract as, for example, a percentage of the final certified value of the contract, a cost per capita, or a percentage of the tender figure or quotation.

Prime costs

The majority of contracts are of such duration that interim payments are sanctioned by the client during construction. The frequency of these payments will depend upon the client's specification for minimum applications, but is usually once a month.

Provided the interim payment is a reliable reflection of the work carried out so far, a comparison can be made between this payment and the expenditure incurred to the same date; this comparison provides an interim picture of the site's profitability.

Site costing

The degree of control that can be achieved by a prime cost is, of course, limited because of its lack of detail. Although a prime cost may expose a frightening loss on a contract, it does not pin-point that loss or indicate its nature. Examples are: loss being made by carpenters on retaining wall structure; loss being made on cement owing to over-rich concrete; loss being made on subsistence owing to insufficient overtime being worked.

To show this kind of detail the contract must have a systematic, regular and continuous check of all the various elements that go to make up the earnings and expenditure of that contract.

The contract's earnings are almost exclusively confined to the amount being paid by the client for carrying out the contract and fall into the following categories.

(1) Measured work paid for at rates agreed between the client and the contractor, either by tender or by negotiation.
(2) Daywork based on rates similarly agreed between the client and the contractor.*
(3) Preliminaries paid for as a percentage or lump sum on the contract figure and similarly agreed between the client and contractor.
(4) Nominated materials and nominated sub-contractors paid for at cost to the contractor, plus an agreed percentage to cover the contractor's obligations of attendance, etc.
(5) Increases in costs where the contract entitles the contractor to recover such increases. These are based on a list of basic prices, which the contractor will have quoted as his basics when tendering, or are based on a formula adjusted according to indices published by the Property Services Agency of the Department of the Environment.

Earnings not paid for by the client are, for example

(1) sundry sales of surplus or scrap materials or plant
(2) sundry works for other clients
(3) tax, or rate refunds, training or government grants.

The contract's expenditure is extremely diverse but can be grouped into the following categories.

(1) The direct cost of labour paid in the form of wages or piecework payments to the contractor's own labour force and to labour-only sub-contractors.
(2) Plant charges, including running costs such as fuel, oil, spares, etc.
(3) Site overheads such as temporary roads, offices, etc., and cost of holiday, insurance stamps, travelling expenses, etc.

* For reference: *Schedule of Basic Plant Charges for Use in Connection with Dayworks under a Building Contract*, The Royal Institute of Chartered Surveyors, 12 Great George Street, Parliament Square, London SW1P 3AD. *The Federation of Civil Engineering Contractors Schedules of Dayworks Carried out Incidental to Contract Work*, Cowdray House, 6 Portugal Street, London WC2A 2HH.

(4) Invoices for materials, both permanent materials such as bricks or cement and temporary materials such as shuttering, sheet piles or scaffolding.
(5) Sub-contractor's accounts based on measured rates or daywork agreed with the sub-contractor.
(6) A proportion of the contractor's head office establishment charges.

This diversity of earnings and expenditure necessitates differing approaches to the various elements to be controlled, but these elements can basically be divided into the following two groups.

Those which do not normally need to be reviewed regularly

(1) materials
(2) sub-contractors
(3) head office charges

Except for the continuous wastage factor which occurs with the material element and the attendance factor which occurs with sub-contractors, the necessity to control this group is normally confined to the initial calculation of

(1) buying margins
(2) sub-letting margins
(3) head office budgeted percentage

Those requiring regular control

(1) labour costs
(2) plant charges
(3) site overheads

These can best be controlled by a system of weekly or even daily costing.

Trade costing

By separating weekly labour, plant and overhead costs on a contract, into trades, a rough outline can be seen of the contract's breakdown of expenditure. If the measure of work done is set against these costs then a rough trade cost per unit of measurement can be calculated and compared with previous week's figures, with other similar contracts or with tender allowances for that trade.

This method of costing, although giving more detail than the monthly prime costs, is clouded by the nature of the work in any particular week and by the variety of items within a trade some of which may be performing better than others.

Unit costing

The calculation of the cost per unit of measurement for each item of work within each trade solves this problem and on a cumulative basis is an ideal form in which to record cost and output data for future estimating purposes as well as providing an element of control required by the site. Such unit costs in isolation, however, may not provide the best impact to prompt corrective action to be taken.

Standard costing

By comparing the costs of each item of work with a known standard value or budget, a picture will emerge of the potential gains and losses on a contract with sufficient clarity and detail to generate immediate action to either reduce expenditure or improve productivity on the individual items; the trade as a whole; the overheads or wherever else the losses have been exposed.

Costing of jobbing works

For small building works, although many of the principles of unit and standard costing still apply, the itinerant nature of the workforce and the shorter duration of the contracts require a more rapid approach without the need for much of the abstracting, analysing and cross checking of records that are necessary on a larger site.

Summary

Many professional occupations have over the years developed their standard methods of working, and staff transfer happily from company to company, even country to country, and are able to pick up the systems used by their colleagues with very little confusion being created. No such standardisation exists in the field of cost control in the construction industry. Study of a company's costing systems will show an ever-changing state of evolution as personalities dictate change of emphasis, market forces affect staff availability and capabilities and, more recently, microcomputers find their way into the industry.

This book, therefore, provides a variety of approaches to costing systems on the basis that, like good clothes, a costing system must be tailor-made to fit a firm, and must be capable of variation to suit particular conditions. Just as a well-dressed man will wear something a little looser for gardening, so a good costing system will discard some of its finer points on a smaller contract, yet the underclothes, the base on which the system is built, remains constant. Otherwise, interchange of staff, collation of data, comparison of problems become a maze of adjusted adjustments that lack the ring of confidence so essential to a good site cost control system.

Chapter 2
Contract accounts and prime costs

Both contract accounts and prime costs require the same type of information, i.e. the total of all earnings and of all expenditure and the comparison of these totals for a contract – at the end of the contract for contract accounts or at intervals during the contract for prime costs.

Example 2A   Contract accounts summary

Earnings	£
Measured work, including variations and PC materials	103 467.90
Dayworks	921.05
Preliminaries	750.00
Nominated sub-contractors	12 325.73
	£117 464.68

Expenditure	£
Workmen's wages	41 228.37
Site staff wages	3 420.91
Labour-only sub-contractors	846.50
Internal plant charges	11 242.06
External plant charges	725.63
Invoices for materials	30 287.82
Payments to sub-contractors (including nominated)	20 423.25
Sundry site expenses	591.62
Head office costs – 2.5% of final certificate, i.e. 2.5% of £117 465	2 936.62
	£111 702.78

Total earnings	117 464.68
Total expenditure	111 702.78
Profit on contract	£5 761.90

$$= 5.2\% \text{ on expenditure}$$
or
$$4.9\% \text{ on earnings}$$

An additional problem that arises with prime costs is the possible earnings to the contract of contentious payments such as revised rates to be negotiated owing to some revision by the client, claims to be discussed or dayworks to be agreed. These items may or may not be paid to the contractor, and it is wise therefore to exclude them from any prime cost calculations showing their *possible* increase to the earnings of the contract after the financial comparison has been made. It is possible that a large contract or even a small problematical contract may carry a heavy sum of contentious items, in which case the amount of contention could be defined under the headings of, for example, 'probable', 'possible' or 'unlikely' payments by the client and a number of differing totals for the contract earnings be obtained. No single formula can be calculated for evaluating contentious sums, and each contract must be considered separately when large contentions are involved. However, the fact remains that, regardless of the amount of the contention, the true earnings of a contract must not include that contention in the initial comparison.

Retention, however, is a different matter, as the client is obliged eventually to pay it to the contractor upon completion of the works. Retention need not therefore be deducted from the earnings in the comparison on the assumption that the contractor will, in fact, complete the contract.

Example 2B Prime cost summary

Earnings	£
Measured work, including variations and PC materials	26 150
Dayworks authorised to date	220
Preliminaries	450
Materials on site	200
Nominated sub-contractors	—
	£27 020

Expenditure	£
Workmen's wages	15 025
Site staff wages	1 200
Labour only sub-contractors	450
Internal plant charges	1 043
External plant charges	185
Invoices for materials	6 700
Delivery notes for invoices not yet received	130
Payments to sub-contractors	715
Sundry site expenses	124
Head office costs – 2.5% of interim certificate, i.e. 2.5% of £27 020	675
	£26 247

Total earnings		27 020
Total expenditure		26 247
Profit on contract		£773

= 2.95% on expenditure

or

2.86% on earnings

Contentious items	£
Dayworks outstanding	53
Variations rejected	28
Possible additional earnings and profit	£81
Total possible earnings	27 101
Total possible profit	£854

$$= 3.25\% \text{ on expenditure}$$
$$\text{or}$$
$$3.15\% \text{ on earnings}$$

If the earnings have been loosely calculated to provide a quick and approximate interim valuation, any comparison between earnings and expenditure can only have a limited credibility. By setting a maximum variation either way to the valuation it may be possible to make limited use of such a figure.

Example 2C   Limit of accuracy of prime cost

Total earnings	£123 500
Possible variation between limits of +3 and −5%	
Therefore maximum value of earnings	£127 205
Minimum value of earnings	£117 325
Total expenditure	£128 135

Therefore loss between limits of £930 and £10 810 = 0.73 to 8.45% on expenditure.

Programmed earnings

By evaluating the bar lines or activities of the contract programme at Bill of Quantity rates, the programmed earnings of the contract can be calculated; these shown as a table or as a graph can be compared with actual earnings and will indicate the financial progress of a contract, as indicated in Figure 2.1.

Prime cost breakdown

By applying percentages to the earnings side of a prime cost it is possible to compare the various elements of expenditure with the same elements of earnings. Rule of thumb, statistics from previous contracts or even a

10 Contract accounts and prime costs

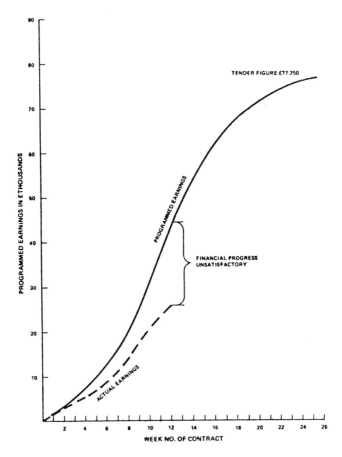

Figure 2.1 Programmed earnings of contract

complete programmed breakdown of Bill of Quantity* rates can be used to calculate these percentages to an acceptable degree of accuracy.

Example 2B shows total earnings of £27 020 which may be broken down as follows

Earnings	Labour			Plant		Materials		Sub-contractors		HO costs	
	%	£	%	£	%	£	%	£	%	£
£200	Material on site					200					
£26 820	60	16 090	15	4020	20	5360	2.5	675	2.5	675
£27 020		16 090		4020		5560		675		675

* The list of items of work in a contract, together with the quantity of each item. The contractor prices the items in the BOQ in order to determine his tender or bid for the contract.

Contract accounts and prime costs

The expenditure is already broken down

Expenditure	Labour	Plant	Materials	Sub-contractors	HO costs
	£	£	£	£	£
Workmen's wages	15 025				
Site staff wages	1 200				
Labour-only contractors	450				
Internal plant charges		1043			
External plant charges		185			
Invoices for materials			6700		
Invoices not yet received			130		
Sub-contractors				715	
Sundry site expenses	124				
Head office costs					675
Total £26 247	16 799	1228	6830	715	675

Comparison of earnings and expenditure shows the following

Comparison	Labour	Plant	Materials	Sub-contractors	HO costs
	£	£	£	£	
Gains	—	2792	—	—	—
Losses	709	—	1265	40	—

This indicates a loss somewhere on the material element and the possibility of work being carried out by hand when machine methods had been envisaged at tender.

Excercise

Calculate a prime cost based on the following information.

	£
Measured work to date	240 390
Variations authorised to date	34 806
Variations in contention	5 723
Nominated sub-contractors	21 240
Dayworks authorised to date	8 916
Dayworks outstanding	721
VOP payments authorised	1 920
Materials on site	400
Preliminaries	7 600
Claim on preliminaries for extended contract	1 800
Labour on loan to other sites	751
Labour on loan from other sites	73
Workmen's wages	103 256
Site staff wages	5 634

12   Contract accounts and prime costs

Invoices for materials	64 220
Delivery tickets for invoices not yet received	350
Sundry site expenses	920
External plant charges	8 220
Internal plant charges	16 575
Sub-contractors (including nominated)	89 372
Head office costs – 4% of site expenditure	

Group studies

(1) Discuss the various problems encountered in obtaining the information necessary to calculate a prime cost.

(2) Discuss the various methods of charging a contract with head office overheads and advise on the most suitable method for companies carrying out various types of work.

Chapter 3
Materials

Although it is possible to allocate materials to individual operations and thus include a material element in any unit or standard costing system, the effort that has to be put into such an exercise does not make such a method practicable for regular costing and should therefore only be used when losses on materials cannot otherwise be identified.

There are two danger points that may cause a loss of money on materials, and a check on these two points is usually sufficient to control material losses.

Buying margin

Normally an estimator will have based his pricing of the material element in a Bill of Quantity rate on a quotation received from a local supplier of each type of material. However, owing to an unrecoverable increase in prices or to the lack of quotation at tender, the amount that the contractor has to pay for his material may eventually be more than he has allowed for. For example: at tender a contractor allows for using concrete aggregate from a local gravel quarry at a quotation of £5.00/tonne, but when he commences construction, he finds that the aggregate is not up to specification and he has to purchase elsewhere at £7.00/tonne. He is therefore left with a £2.00 loss on every tonne of aggregate. If the estimated amount of aggregate in all concrete mixes for the contract is 1000 tonne, it follows that a loss of £2000 is inevitable on this material.

This may be offset by gains on other materials where, for instance, a post-tender quotation has shown a saving on the material element used by the estimators.

Such gains and losses on the buying margin of all materials can be tabulated as illustrated in Figure 3.1 in order to calculate the overall buying margin for a complete contract, the material element being the tender value for material excluding planned profit but including any margins allowed by the estimators for increased rates, etc.

Only if constant non-recoverable alterations to material prices continue throughout a contract is it necessary to review this calculation regularly.

ANY FIRM & CO. LTD.				CONTRACT		
MATERIAL BUYING MARGINS						
MATERIAL	VALUE ALLOWED PER UNIT £	ACTUAL COST PER UNIT £	GAIN PER UNIT £	TOTAL GAIN £	LOSS PER UNIT £	TOTAL LOSS £
5000 tonnes hardcore	3.00	2.25	0.75	3750	–	–
600 tonnes cement	50.00	50.00	–	–	–	–
100 tonnes building sand	4.00	4.50	–	–	0.50	50
1500 tonnes concreting sand	5.00	5.50	–	–	0.50	750
4000 tonnes aggregates	7.00	6.00	1.00	4000	–	–
120 thousand common bricks	45.00	48.00	–	–	3.00	360
30 thousand engineering bricks	60.00	70.00	–	–	10.00	30

	7750	1190
	1190	
OVERALL BUYING MARGIN:	£6560	

Figure 3.1 Calculation of buying margins on materials

Reference to the buying margin must always be made before placing emergency orders with an alternative supplier owing to difficulty in obtaining materials, and the advantage of speeding up deliveries must be balanced with the increase in costs. Where the same material is being purchased from two or more suppliers, the cheaper supplier must be clearly marked so that daily or weekly requirements do not go to the dearer supplier simply because he arrives a little more quickly or provides a marginally better service.

Wastage

Inevitably a certain percentage of wastage must be incurred on any construction site owing to cut-offs in timber, snap headers and closers in brickwork, consolidation in hardcore, etc. Allowances will have been

CONTRACT _____   DATE up to 9/5/82

MATERIAL	UNIT	VALUE			ACTUAL			TO DATE			SINCE LAST CHECK		
		QUANTITY PLACED	WASTAGE ALLOWANCE	TOTAL ACCOUNTED FOR	BOUGHT	STOCK	USED	GAIN	LOSS	%	GAIN	LOSS	%
Bricks	No.	358 x 116 41 528	5% 2 075	43 603	53 000	1 000	52 000		8 400	20%			
100 mm dia. pipes	m	90	5% 5	95	125	30	95	—	—	—			
150 mm dia. pipes	m	300	15	315	420	90	330	—	15	5%			
Mesh reinforcement	m²	140	15% 21	161	210	60	150	11	—	8%			

Figure 3.2 Calculation of wastage on materials

made at tender for such wastage but must be checked to ensure that normal limits are not being exceeded. The amount of each material purchased must be compared with the amount accounted for.

Figure 3.2 calculates the wastage for a sample of four items of material based on the following information

Deliveries to site

Bricks	53 000
100 mm diameter pipes	125 m
150 mm diameter pipes	420 m
Mesh reinforcement	210 m^2

Stock on site

Bricks	1 000
100 mm diameter pipes	30 m
150 mm diameter pipes	90 m
Mesh reinforcement	60 m^2

Amount in measured work

Brickwork 1B thick	358 m^2
100 mm diameter pipes	90 m
150 mm diameter pipes	300 m
Mesh reinforcement	140 m^2

Waste

Allow 5% planned waste on bricks and pipes
Allow 15% planned waste and laps on reinforcement

Care must be taken to allow for all purchases. The addition of delivery tickets is acceptable provided that strict control is observed over handing these tickets into the office or a *goods received* book is kept. Otherwise, invoices should be used for calculating deliveries, an allowance being made from delivery tickets for materials not yet invoiced. Where alternative suppliers are used in an emergency, care must be taken to include these suppliers even though their contribution may be small.

Stock on site is sometimes difficult to calculate exactly but as the work progresses, the amount of materials stored on site will reduce thus making the comparison more accurate. Stocks are, of course, more easily estimated when they are low: early Monday morning, after the site has worked for the weekend, may thus prove the ideal time to take stock of, for instance, concreting materials. Stacked bricks can be calculated at 500 per cubic metre downwards, depending on how loosely stacked they are, and cement silos can be measured more accurately when they are full. In order to reduce the effect of inaccurate estimates of stock, these checks are better calculated to date; weekly or monthly figures can then be deducted and shown relative to the *to-date* results; this also helps to indicate trends of wastage during the course of the contract.

Records of temporary works on site should be included in order that all materials used can be allowed for: for example, concrete in base slab for site office, drain pipes to temporary toilets, etc.

It is often found that the method of calculating concrete mixes can be the cause of a gain or loss on concreting materials; if it is discovered that the rule of thumb method used at the estimating stage is inaccurate in practice, then the comparison for wastage purposes is best based on a realistic concrete mix. Any gains or losses caused by the estimator's method of assessing quantities of material in a concrete mix can be calculated as a variation of the buying margin. Otherwise, actual amounts used would be compared with unrealistic standards.

Exercise

Show how you would set up a system of control for wastage and buying margins of concreting materials. Assume that the estimate has been based on a rule of thumb method of calculating the mix, whereas on site concrete is weighbatched. Assume also that cheaper quotations have been obtained since tendering.

Chapter 4

Sub-contractors

Letting margin

As with materials, it is necessary to check buying margins or letting margins on domestic sub-contractors at the commencement of a contract. Nominated sub-contracts are selected by the client and cannot be varied by the contractor; thus such a comparison is unnecessary except as a service to the client in obtaining quotations from various nominated firms.

ANY FIRM & SON LTD.			CONTRACT	
SUB-CONTRACTORS LETTING MARGINS				
TYPE OF WORK	VALUE OF SUB-CONTRACT IN TENDER £	ACTUAL SUB-CONTRACT £	GAIN £	LOSS £
Structural Steelwork	25 340	25 340	–	–
Plumber	10 961	10 400	561	–
Plasterer	4 210	4 530	–	320
Felt Roofer	2 200	1 950	250	–
Painter	1 083	1 083	–	–
Fencer	846	813	33	–
		Overall Letting Margin	844 320 £524	

Figure 4.1 Calculation of letting margins on sub-contractors

Figure 4.1 shows the comparison of domestic sub-contracts, which would not normally require to be revised during the course of a contract unless a sub-contract was changed or the sub-contractor's quotation altered owing, for instance, to inability to achieve programme or alteration in conditions on site beyond the control of the sub-contractor.

Attendance

Few contracts are completed without the main contractor providing some service or attendance on the sub-contractor. This may be recoverable from the client either as a daywork or as a measurable item or may be recoverable from the sub-contractor (depending on the conditions of sub-contract) as a deduction from his account. Such attendance may, however, be the responsibility of the main contractor and as such must be included in the site costing system so that a regular comparison can be carried out with whatever allowances were made at tender. This can be done as a fixed-labour overhead, the value being either a fixed sum per week or related to the amount of sub-contractor's work completed. As sub-contractors are usually valued only monthly, a conversion of a percentage allowance to a weekly allowance is necessary for a weekly costing system. Probably the simplest method of budgeting for attendance on sub-contractors is to allow a lump sum standard which is drawn off as the sub-contractor's work proceeds in a similar manner to that described in Chapter 10 in connection with temporary roads.

Chapter 5

Trade costs

The principle of *trade costs* is a simple one; all labour and plant expenditure by each trade must be allocated against that trade and then divided by the measure of work done during the same period. For example, Figure 5.1 indicates that during the week ending 22 August, 1982 the sum of £250 was spent on labour placing 50 m³ of ready-mixed concrete plus £125 spent on the hire of associated plant. The trade cost of placing concrete therefore is

$$£250 + £125 \div 50\,\text{m}^3 = \underline{£7.50/\text{m}^3}$$

In order to compare this figure with previous and future weeks, or with costs to date, or other contracts or tender allowances, the extent of the labour and of the plant that is to be included in each trade must be clearly defined. It is misleading if one contract is, for instance, including in its cost of laying brickwork the mixing of mortar, whilst another excludes such costs, or one site is including fuel costs for plant whilst another bases its costs on hire rates only. It is not comparing like with like if one week the measure of formwork includes stop ends but excludes such items the following week or if one contract is measuring formwork when fixed and another when stripped.

Equally the items embraced by each trade must be carefully defined as must the separation of productive work from overhead costs. It is not compatible if one site is charging scabbling construction joints against the concrete trade when another is allocating this operation to formwork or if one site considers the employers' contribution to the Graduated National Insurance scheme to be an overhead whilst others include it as part of the construction costs. Thus, what started as a simple principle requires a complex procedure in order to obtain results that will withstand comparison.

A suggested format for presentation of the trade costs is shown in Figures 5.1 and 5.2, the headings being defined as follows.

Productive work

The broad headings or trades against which costs are to be allocated, e.g. bulk earthmoving, place concrete, formwork, etc. Each heading will need to be defined as suggested by the list of definitions shown in Figure 5.3 for external works, Figure 5.4 for structures and Figure 5.5 for house construction. Special items within any trade can be shown as a separate heading. These fall into a number of groups as follows.

(1) Specialist work, e.g. de-watering, concrete runways, chimney slip forming, kiln linings, etc.
(2) Work unusual to the nature of the job, e.g. underpasses on a housing contract, building work on a road contract, etc.
(3) Work normally carried out by a sub-contractor, e.g. floor screeds, paramount partitioning, painting, road markings, etc.
(4) Major items of remedial work, e.g. wind damage, flood damage, etc.
(5) Sectionalisation of the contract into different buildings, structures, etc.

Each of these items can be costed separately though too much subdivision can lead to the downfall of the system which is designed with simplicity in mind. A good yardstick is to aim to a restrict the number of headings to only those that can be entered on a single sheet of A4 paper even if this means grouping some operations together. This is illustrated in Figure 5.2 where some of the headings for road works suggested in Figure 5.3 have been grouped together under the heading of Trim and Lay Roads. Clearly the items included in the grouping need to be defined to show, for example, that black top is sub-contracted and therefore not included.

All labours and sundry items associated with each trade are to be included with that trade. This covers any making good, removal of spillage and cleaning up on completion – except for the final snagging – and cleaning of site prior to handover which is usually difficult to apportion against individual trades. Snagging to housing warrants a separate productive work heading, whereas the final cleaning of site has been included under the definition for site attendances.

Site overheads work

Before embarking on any cost system decision must be made as to where the line is drawn between *productive work* and work that is to be classed as a *site overhead*. Clearly, the setting up of *site facilities* – office, canteen, stores – must be treated as an overhead but the handling of materials – unloading, transporting and hoisting – could be argued to be either productive or overhead in nature. A company's established estimating system will probably decide this argument but from an allocation of costs viewpoint it is relatively difficult to correctly allocate the hours of labour and plant that are handling materials against each of the different trades including apportioning the time when that labour or plant is standing, e.g. crane and driver at an isolated bridge construction, material handling gang on a housing site, hoist and driver on a multi-storey office block. It is by no means impossible to obtain allocation from such situations but the end may not justify the means and may, if attempted only half heartedly, contaminate the otherwise excellent records of costs expended against each trade heading.

PRODUCTIVE WORK	Budget £	THIS WEEK							TO DATE				
		Measure	Labour hours	Output h/unit	Labour	Plant	Total	£/unit	Measure	Labour hours	Output h/unit	Labour	Plant	Total	£/unit
EXTERNAL WORKS															
Bulk earthmoving	£1.00/m³								18702	1176		4723	8812	13535	0.72
Main drainage	£6.00/m								980	432		1711	3025	4736	4.83
Branch drains	£9.00/m	12	15	1.2	55	72	127	10.58	72	58		225	380	605	8.40
Manholes	£50.00/m								15	108	7.2	423	102	525	35.00
STRUCTURES															
Excavation	£9.00/m³								624	183		692	880	1572	2.52
Place concrete	£15.00/m²	50	68	1.4	250	125	375	7.50	733	915	1.2	3281	1425	4706	6.42
Make, fix & strip formwork	£300/set	72	158	2.2	681	152	833	11.57	918	1722	1.9	7161	1094	8255	8.99
Fix holding down bolts	£10.00/1000	4	13	3.2	51	16	67	16.75	30	144	4.8	515	138	713	23.76
Mixe & transport mortar	£120/1000	10,400	45	4.3	182	120	302	29.04	10,400	61	5.9	235	160	395	37.98
Lay bricks and blocks		10,400	232	45/hour	1155	38	1193	114.71	10,400	232	45/hour	1155	38	1193	114.71
LABOUR-ONLY SUB-CONTRACTORS															
Fix reinforcement	£80/tonne	3	75	25	353	10	363	121.00	38	690	18	3250	112	3362	88.47
TOTAL PRODUCTIVE WORK			606		2727	533	3260			5721		23431	16166	39597	

CONTRACT ANY Factory Site   TRADE COSTS   Week No 20 of 52 W/E 22 Aug 82

SITE OVERHEADS WORK		Budget £ %	Labour hours	% hrs on prod've	THIS WEEK £ COST			% £ on prod've	Labour hours	% hrs on prod've	TO DATE £ COST			% £ on prod've
					Labour	Plant	Total				Labour	Plant	Total	
Material handling	Unload materials	1%	12	2.0%	45	10	55	1.7%	68	1.2%	259	64	323	0.8%
	Transport on site	5%	21	3.5%	79	35	114	3.5%	113	2.0%	448	212	660	1.7%
	Cranage and hoists	1%							55	1.0%	216	308	524	1.3%
Plant maintenance		2%	25	4.1%	108	8	116	3.6%	283	4.9%	1152	92	1244	3.1%
Site facilities		5%				65	65	2.0%	138	2.4%	502	1522	2024	5.1%
Site attendances		10%	68	11.2%	212	104	316	9.7%	818	14.3%	2463	1012	3475	8.8%
Attend sub-contractors		1%	5	0.8%	20	14	34	1.0%	55	1.0%	213	82	295	0.7%
Staff foremen														
Dayworks														
Total site overheads work		25%	131	21.6%	464	236	700	21.5%	1530	26.7%	5253	3292	8345	21.1%
Productive work brought forward			606		2727	533	3260		5721		23431	16166	39597	
TOTAL ALL WORK			737 a		3191 b	769 x	3960		7251 c		28684 d	19458 y	49742	

BALANCE OF LABOUR	THIS WEEK		TO DATE		
	Hours	£ cost	Hours	£ cost	
Hourly paid	630	2758	6007	23772	
Redundancy rebate					
Shift & weekly	32	80	364	910	
Works staff					
Hired drivers					
Lab only subs	75	353	690	3250	
+ On loan in			230	912	
− On loan out			−40	−160	
	737 a	3191 b	7251 c	28684 d	

BALANCE OF PLANT	WEEK	TO DATE
	£ cost	£ cost
Company heavy plant	392	10382
Hired heavy plant	58	340
Company light plant	89	803
Hired light plant	20	120
Plant consumables	210	7725
+ On loan in		88
− On loan out		
	769 x	19458 y

COMMENTS

Prepared by _____

Figure 5.1 Trade cost for a general building contract

PRODUCTIVE WORK	Budget £	THIS WEEK							TO DATE						
		Measure	Labour hours	Output h/unit	Labour	Plant	Total	£/unit	Measure	Labour hours	Output h/unit	Labour	Plant	Total	£/unit
Bulk earthmoving	1·00/m³								62540	5208		20714	35725	56439	0·90
Main drainage	7·00/m								2610	1525		6128	8016	14144	5·42
Branch drains	9·00/m	20	353		121	48	169	8·45	683	638		2563	1987	4550	6·66
Manholes	60·00/m								112	1320	11·8	5180	920	6100	54·46
Road gullies	25·00/No								72	308	4·3	1212	313	1525	21·18
Trim & lay roads	2·00/m²								6320	3153	0·50	12504	9227	21731	3·44
Curbs	4·00/m								2230	1705	0·76	7120	132	7252	3·25
Paths and paved areas	2·00/m²	42	20	0·48	83	—	83	1·98	2157	830	0·38	3398	216	3614	1·68
Fencing	5·00/m	150	138	0·9	568	42	610	4·07	262	283	1·1	1180	84	1264	4·82
TOTAL EXTERNAL WORKS	1700/No	0·5	191	382	772	90	862	1724	72	14970	2·08	59999	56620	116619	1620
Excavate strip founds	60/No								80	868	11	3380	869	4249	53
Concrete strip founds	55/No								80	742	9	2950	320	3270	41
Stone slab (inc trim)	72/No	2	52	26	228	43	271	136	78	1487	19	5846	1011	6857	88
Concrete slab	70/No								72	1233	17	4936	513	5449	76
Brickwork to foundations	150/No	3	112	37	488	5	493	164	75	2455	33	10283	233	10516	140
Brickwork to super	1000/No	3	718	239	3092	24	3116	1039	52	12360	238	56230	715	56945	1095
Mortar supply	180/No	3	98	33	406	223	629	210	56	1653	30	6562	3038	9600	171
Carcassing	160/No	6	223	37	912	5	917	153	44	1890	43	7666	53	7719	175
First fix joinery	200/No	5	245	49	1001	38	1039	208	32	1443	45	5972	218	6190	193
Second fix (inc ironmongery)	200/No	5	202	40	818	30	848	170	18	763	42	3511	134	3645	203
Garages	120/No								20	632	32	2568	115	2683	134
Scaffolding	320/No	4	172	43	701	683	1384	346	44	1971	45	8033	7256	15289	347
Snagging	50/No	4	112	28	486	31	517	129	10	186	19	750	68	818	82
Maintenance period	(20/No)														
TOTAL HOUSE CONSTRUCTION	2637/No		1934	532	8132	1082	9214	2555		27683	583	118687	14543	133230	2798
TOTAL PRODUCTIVE WORK	4337/No		2125	914	8904	1172	10076	4279		42653	791	178686	71163	249849	4418

CONTRACT ANY Housing Site   TRADE COSTS   Week No 50 of 78 W/E 9 May 82

SITE OVERHEADS WORK		Budget £ %	THIS WEEK					TO DATE				
			Labour hours	% hrs on prod've	£ COST			Labour hours	% hrs on prod've	£ COST		% £ on prod've
					Labour	Plant	Total	% £ on prod've			Labour	Plant	Total	
Material handling	Unload materials	1%	20	0.9%	82	–	82	0.8%	947	2.2%	3813	143	3956	1.6%
	Transport on site	7%	90	4.2%	376	207	583	5.8%	4113	9.6%	15845	8292	24137	9.7%
	Crane	NIL							153	0.4%	613	1420	2033	0.8%
Plant maintenance		1%	30	1.4%	121	25	146	1.5%	501	1.2%	1998	173	2171	0.9%
Site facilities		5%				481	481	4.8%	1016	2.4%	4121	11284	15405	6.2%
Site attendances		10%	169	8.0%	612	189	811	8.0%	5625	13.2%	20820	6222	27042	10.8%
Attend sub-contractors		1%	88	4.1%	180	10	190	1.9%	483	1.1%	1920	104	2024	0.8%
Staff foremen														
Day works														
Total site overheads work		25%	397	18.7%	1371	912	2293	22.8%	12838	30.1%	49130	27638	76768	30.7%
Productive work brought forward		4357	2125	914	8904	1172	10076	42.9%	42653	791	178686	71163	249849	44.18
TOTAL ALL WORK		5421	2522 a	1085	10275 b	2084 x	12369	5255	54491 c	1029	227816 d	98801 Y	326617	57.74

BALANCE OF LABOUR	THIS WEEK		TO DATE	
	Hours	£ cost	Hours	£ cost
Hourly paid	2492	10195	53920	225771
Redundancy rebate				
Shift & weekly	30	80	260	790
Works staff				
Hired drivers			153	613
Lab only subs				
+ On loan in			198	812
– On loan out			– 40	– 170
	2522 a	10275 b	54491 c	227816 d

BALANCE OF PLANT	WEEK		TO DATE	
	£ cost		£ cost	
Company heavy plant	1016		5325	
Hired heavy plant	108		1420	
Company light plant	126		4265	
Hired light plant	11		220	
Plant consumables	823		3894	
+ On loan in				
– On loan out				
	2084 x		9880 1 Y	

COMMENTS

Prepared by _____

Figure 5.2 Trade cost for a housing contract

	Measure
(1) Demolition and Site Clearance	
Initial site clearance such as grubbing up hedges, trees, etc., and the demolition of existing buildings	Total
(2) Bulk Earthmoving	
All bulk reduce level excavation sub-divided where possible by differing machines, by differing haul lengths and by excavation or fill in differing materials	m^2
(3) Main Drainage	
Excavation, pumping, planking and strutting, placing concrete, pipe laying, sanding, stoning and backfilling to all main drainage runs	m
(4) Branch Drains	
All drainage as above but to branch drains in connections to housing or buildings or cut-off drains in road embankments	m
(5) Manholes and Catch Pits	
The complete construction of manholes and catch pits up to and including the laying of covers and frames and including any excavation carried out separate from the drain run excavation	m depth
(6) Road Gullies	
The complete construction of road gullies up to and including the laying of frames and grids and including any excavation carried out separate from the drain run excavation	No.
(7) Services	
Excavation, pumping, planking and strutting, placing concrete, pipe and duct laying, sanding, laying of PC cable covers and backfilling to all gas, telephone, water and electricity services	m
(8) Trimming Road Formation	
Final trimming of road formation (second stage excavation)	m^2
(9) Stone Sub-Base to Roads	
Laying and levelling of stone sub-base to roads	m^2

	Measure
(10) <u>Stone Base Course to Roads</u>	
Laying and levelling of each layer of stone base course on top of sub-base	m²
(11) <u>Rigid Paving</u>	
All shuttering, reinforcement and concrete involved in concrete road or similar concrete area	m²
(12) <u>Flexible Paving</u>	
All tarmacadam (black top) paving sub-divided into base course and wearing course	m²
(13) <u>Kerbs</u>	
Shuttering and concrete to kerb race, laying and jointing kerbs and haunching to back of kerb stones	m
(14) <u>Cats Eyes</u>	
Breaking out or drilling and inserting cats eyes in roads or paved areas	No.
(15) <u>Paths and Pavement Areas</u>	
Excavating, laying, levelling and rolling sand, ash, etc., and laying and jointing of paths and pavings	m²
(16) <u>Edgings</u>	
PC or timber edgings to paths and paved areas	m
(17) <u>Fencing</u>	
Excavating, positioning, driving, placing of concrete, backfilling, erecting and painting of fence posts and fencing	m
(18) <u>Topsoiling</u>	
Spreading and laying of topsoil	m²
(19) <u>Grassing</u>	
Laying of turf or grass seed	m²
(20) <u>Total External Works</u>	
Totals for all stages of external works construction	

Figure 5.3 Definition of trade headings for productive work (external work)

	Measure

(1) Excavation

All excavations to foundations including pumping and planking and strutting — m³

(2) Backfilling

All backfilling with excavated material or imported fill or stone — m³

(3) Cofferdams

All sheet piling and cofferdams — m²

(4) Piling

Driving, re-driving withdrawing shells and cutting off pre-cast concrete piles — m penetration

(5) Supply Concrete

All work involved in the setting up and dismantling of batching plant and in the mixing and transporting of concrete up to the point that 'ready-mix' concrete would be supplied, i.e. local to the place of construction. Mixing of dry lean concrete to be separated from concrete. Measure of concrete to be that subsequently placed — m³

(6) Place Concrete

Placing of concrete and all labours such as vibrating, curing, trowelling, filling bolt holes, rubbing up and scabbling. All concrete to be included unless shown as part of a separate heading, e.g. laying dry lean to roads, concreting kerb races, benching manholes, etc. — m³

(7) Special Concrete Items

Substantial items associated with concreting that would unreasonably distort the concrete output if included in that trade, e.g. extensive decorative bush hammering, cement and sand screeds, monolithic grano, mastic pointing, saw cutting joints, etc. Each special item to be kept separate — as appropriate

(8) Formwork

Make, fix and strip including standby on concrete pours, oiling shutters, repairing and recovering materials. All shuttering actually carried out to be measured when fixed including stop-ends but only for the surface area of the concrete regardless of shutter size. Edges of slabs, strings and risers to stairs, beams and plinths to be converted to m² and included. Hours spent on all sundry labours and small enumerated and lineal items, kickers, keys, channels, holes, pockets, slots, bolts, dowels, pads, unistruts, abbeyslots, etc., to be included with no additional measure. Hours spent on erecting trishors and acrowprops are to be included — m² fixed

(9) Special Formwork Items

Substantial items associated with formwork but which would unreasonably distort the formwork output if included in that trade, e.g. QC decking, extensive featured surfaces, sets of holding down bolts, large areas of flexcell, polystyrene, etc. Each special item to be kept separate — as appropriate

(10) Cut and Bend Rod Reinforcement

Cutting and bending of rod reinforcement and all sorting for cutting and bending, labelling, storing and removal of off cuts — tonnes

(11) Fixing Rod Reinforcement

Fixing and tieing of rod reinforcement including any wire brushing, straightening, re-fixing and all sorting for fixing. All site making and fixing of spacer blocks to be included — tonnes

	Measure

(12) Mesh Reinforcement

Sorting, cutting, bending and fixing of fabric reinforcement — m^2

(13) Special Reinforcement Items

Substantial items associated with reinforcement but which would unreasonably distort the cutting and bending or fixing output if included in that trade, e.g. pre-stressing, welding, etc. Each special item to be kept separate — as appropriate

(14) Brickwork and Blockwork

All brick and blockwork including corbels, arches, manholes, sleeper and honeycombed walls. Half brick and split courses to be measured as part bricks only. Hours spent on pointing, forming piers or openings, cutting and pinning to soffites, etc. and all sundry labours and lineal and enumerated items, e.g. cills, lintols, air bricks, DPC, brick reinforcement, cavity filling, door templates, etc., to be included with no additional measure. Blocks to be converted to bricks at 30 bricks per m^2 of 100 mm wide blocks. Mixing and transporting of mortar to be shown separately — 1000 No.

(15) Mortar Supply

All work involved in the setting up and dismantling of the mortar mixing plant and in the mixing and transporting of mortar up to the point that 'ready-mix' mortar would be supplied, i.e. local to the place of construction. For measurement purposes blocks are to be converted to bricks as above — 1000 No.

(16) Special Brick and Blockwork Items

Substantial items associated with brick or blockwork which would unreasonably distort the brick or blockwork outputs if included in that trade, e.g. floor tiling and screeds, extensive cutting holes and chases for services, special brick or uni-block paving, brick channels, cobblestones, etc. — as appropriate

(17) Pre-Cast Units

Hoisting, erecting, bolting, temporary supports, jointing and cleaning to be included. Craneage used exclusively for erection as opposed to general material handling to be included. Small items of lintols, cills, etc., normally erected with bricks/blockwork not to be included — tonnes

(18) Structural Steel

Hoisting, erecting, bolting, welding, temporary supports, cleats, etc., and all cleaning to be included. Craneage used exclusively for erection as opposed to general material handling to be included — tonnes

(19) Carpenter and Joiner

Carcassing, first and second fixings and ironmongery — total

(20) Scaffolding

Cutting, sorting, erecting and dismantling scaffold tubes, boards and fittings including setting out, positioning and levelling base plates. All types of scaffolding to be included, barrow runs, independent and putlog scaffold, birdcage towers, cantilevers, canopies, chutes, bracings, safety notices and boards, toe-boards, ladders, etc. Trishores and acroprops to formwork are not to be included under scaffolding. Conventional tubes and fittings to be kept separate from patent scaffolding — m erected

(21) Total Structures Construction

Totals for all stages of construction of the structures

(22) Total Productive Work

Combined totals of external works and structures construction

Figure 5.4 Definition of trade headings for productive work (structures)

	Measure
(1) Excavate Strip Foundations Excavation of foundation trenches including all planking and strutting and pumping	m^3
(2) Concrete Strip Foundations Placing of concrete to foundation trenches including any shuttering, bar or mesh reinforcement	m^3
(3) Concrete or Blockwork Ring Beam to Timber Framed Buildings Shuttering, reinforcement and concrete or laying concrete blocks to ring beams for timber framed buildings	m^3
(4) Excavate Foundation Slab Excavate or trim foundation for foundation slab or raft	m^2
(5) Stone and Prepare Slab Preparation, laying, blinding and rolling of stone to foundation slab	m^2
(6) Concrete Slab Placing of concrete to foundation slabs. Any bar or mesh reinforcement, shuttering, forming or placing of ducts, holes, pockets, bolts, etc., in the slab to be included	m^3
(7) Brickwork to Foundations All brickwork and blockwork up to and including DPC level to be included. Blocks to be converted to bricks at 30 bricks per m^2 of 100 mm wide blocks. Mixing and transporting of mortar to be shown separately	1000 No.
(8) Brickwork and Blockwork to Superstructure All brickwork and blockwork above DPC level including associated sundry items such as pre-cast concrete lintols and cills, brick reinforcement, cavity ties, cavity filling, door templates and forming holes for services. Blocks to be converted to bricks at 30 bricks per m^2 of 100 mm wide blocks	1000 No.
(9) Mortar Supply To include all work involved in the setting up and dismantling of the mortar mixing plant and in the mixing and transporting of mortar up to the point that 'ready-mix' mortar would be supplied, i.e. local to the place of construction. For measurement purposes, blocks are to be converted to bricks as above	1000 No.
(10) Frame Erection to Timber-Framed Buildings Erection of floor and wall panels, internal stud partitions and roof trusses of timber framed buildings including sole plates, fibre-glass installation and any additional panels necessary at staggers in the line of the buildings and including filling and sanding to floors. Craneage used exclusively for erection as opposed to general material handling to be included	dwelling
(11) Pre-Cast Concrete Structures Erection of pre-cast concrete floor, roof or frame units including all bedding, fixing, bolting and jointing and temporary supports. Craneage used exclusively for erection as opposed to general material handling to be included	dwelling

	Measure
(12) Carcassing	
Cutting, framing, erection and fixing of joists and roof timbers including tank bearers, noggins, barge boards, eaves soffites and fascias	dwelling
(13) First-Fix Joinery	
Cutting, framing, erection, fixing of flooring, door and window frames and boards, studding, pipe casings and staircases	dwelling
(14) Second-Fix Joinery	
Cutting, framing, erection, fixing and hanging of doors, skirtings, architraves, kitchen fittings, shelving and general completion of the carpentry and joinery works	dwelling
(15) Loft Insulation	
Fibre-glass insulation to lofts	dwelling
(16) Ironmongery	
Fixing of ironmongery to doors, windows, cupboards, etc.	dwelling
(17) Porches	
All work in connection with the erection of porches where separable from main dwelling construction	porch
(18) Stores	
All work in connection with stores where separable from main dwelling construction	store
(19) Garages	
All work in connection with garages where separable from main dwelling construction	garage
(20) Scaffolding	
Erection and dismantling of all scaffolding including toe-boards, handrails, ladders and safety barriers	m tube
(21) Snagging	
Snagging of dwellings prior to handover but not to include the time spent on the initial fixing of any items of work constructed out of phase. Where possible snagging by different trades is to be kept separate	dwelling
(22) Maintenance Period	
Making good during the contract maintenance period	dwelling
(23) Total House Construction	
Totals for all stages of construction to houses, flats or bungalows, etc.	dwelling
(24) Total Productive Work	
Combined totals of external works and house construction	dwelling

Figure 5.5 Definition of trade headings for productive work (house construction)

	Measure

(1) Material Handling

(a) Unload – all costs involved in unloading and stacking of plant and materials including cranes and fork lift trucks involved in unloading — Percentage

(b) Transport – to include all sorting, labelling, reloading, transporting, hoisting and re-unloading of plant and materials up to the point of construction including general service lorries, water bowsers and dumpers.

All sorting and handling of materials at the place of construction is to be included in the appropriate productive trade.

Cleaning of materials and the removal of scrap, waste or spillage is not to be included under material handling but is to booked against the appropriate productive work — Percentage

(c) Craneage – to include only cranes and hoists and their drivers handling plant and materials at the place of construction. Craneage working exclusively for a production gang is to be allocated against the appropriate productive trade, e.g. pile driving, house frame erection. Banksmen to cranes are to be allocated against the appropriate trade heading — Percentage

(2) Plant Maintenance

Plant maintenance labour and plant including fitters, greasers and fuel bowsers. The cost of drivers *greasing time* is to be included with the appropriate driver and not extracted against plant maintenance — Percentage

(3) Temporary Facilities and Attendances

(a) Site facilities – the erection, hire, general maintenance and dismantling of the following temporary facilities

    Contractor's offices and offices for client and his representatives
    Steel yard, joiners' shop stores and compound but not including productive plant, e.g. bar bender, saw, etc.
    Ablutions, drying and messing accommodation
    Diversions, hardstandings, car parks and temporary bridges
    Crane rails, accesses, fencing and notice boards
    Traffic signals, watching, safety and security precautions
    Labour camp and canteens
    Services to any of the above — Percentage

(b) Site attendances – attendance on the above site facilities as follows

    Vans, cars, minibuses and coaches and their drivers
    Storemen, checkers, chainmen, laboratory assistants, concrete testing assistants and office staff paid on the workmen's payroll
    Cleaners, tea women, toilet and office attendants
    Labour camp or caravan park attendants
    Watchmen, security and first-aid attendants
    Road sweepers and cleaning of public or client's roads
    Shop Steward's and Convenor's duties full or part-time
    Final cleaning site, drying and cleaning out buildings — Percentage

(4) Attendance on Sub-Contractors

Attendance on both nominated and domestic sub-contractors but not including attendance on labour-only sub-contractors which should be shown against the appropriate trade — Percentage

(5) Daywork

Where costs recoverable on daywork are substantial the amounts actually recovered from the client are to be extracted from the trades. If a large amount of daywork is in contention then the to-date total is to be adjusted as the contention is resolved

(6) Foreman on Workmen's Payroll

The cost of any supervisory foremen or walking gangers, i.e. in charge of more than one gang of men who are paid on the workmen's payroll — Percentage

(7) Total Site Overheads Work

The total of all hours and costs listed above expressed as a percentage on the total hours and costs of productive work — Percentage

Figure 5.6 Definition of trade headings for site overheads work

It is therefore preferable to consider *material handling* as a site overhead and, together with other overheads, monitor its cost as a percentage on productive work. The dividing line between assisting in the productive work and the general material handling on site can be taken as the point on the scaffold or area of the work where the general material handling gang, crane, hoist, conveyor, fork lift truck or other means of transport, delivers the materials local to the area of construction.

Figure 5.6 gives a suggested list of definitions for site overheads work but, as with the productive work, this list can be further broken down to record, for instance, attendances against individual sub-contractors, cost of site facilities and site attendances for the client separate from those of the contractor, the cost of apprentices or the cost of fuel and plant spares instead of including these under productive work. The trade cost can be extended to include such expenditure as staff costs, insurances, stationery and telephone costs, etc.

Budgets

The budget for each productive work heading is to be the estimated or target cost of that trade or item expressed either as a total budget per week such as place concrete £150, or as a budget per unit of measurement, place concrete £9.00/m^3 indicated in Figure 5.1.

On *housing contracts* an alternative is to express the budget as an amount per dwelling, e.g. second-fix joinery £200 per dwelling. This enables the budget for all stages of construction to be totalled to give a budget per dwelling up to any point of construction. As illustrated in Figure 5.2, any stages that have not yet commenced are not to be added into this total budget.

The budget for each site overhead can also be expressed as a total budget per week such as site attendances £150, or as a budget percentage on either the hours or the cost of the productive work, e.g. site attendances budget of 10% on the cost of productive work indicated in Figure 5.1.

Where the cost records are to be taken as costs current at time of construction then any national increases that are recoverable under the contract or that have been allowed for in the contractor's tender should be taken into account when setting such budgets and should be reviewed when substantial national increases occur during the course of the contract. The notes on national increases (see Chapter 24) discuss this point more fully.

Measure

The method of measurement under any particular trade heading must always remain constant regardless of how individual items are purchased or how they are shown in the Bill of Quantities or in the appropriate Standard Method of Measurement.*

---

* For reference: *Standard Method of Measurement of Building Works*, authorised by agreement between the Royal Institute of Chartered Surveyors and The National Federation of Building Trade Employers. *Civil Engineering Method of Measurement*, sponsored by The Institution of Civil Engineers and The Federation of Civil Engineering Contractors.

Figure 5.3 suggests suitable units of measurement for each trade heading for external works or road contracts and Figure 5.4 for buildings or structures. Figure 5.5 suggests suitable units of measurement for housing contracts, however, an additional set of trade costs for housing can be obtained by taking a count of the number of dwellings completed for each of all the stages of construction including an assessment for partly completed dwellings in any of the stages. When a dwelling does not have a particular stage – garages, porches, etc. or when work is not directly related to the number of dwellings – main drainage, roads, etc., an assessment has to be made of the percentage of that item completed within the total quantity for the site and this expressed as a percentage of the total dwellings. For example ten garages are complete on a housing site containing 40 garages, so garages are 25% complete. If the contract is for 80 dwellings then the measure for garages would be 25% of 80, i.e. 20 dwellings. By the completion of the contract, measures in all stages will have reached 80.

The advantage of measuring housing in this way is that the total number of man hours per dwelling and the total cost per dwelling will start to be apparent the moment the first dwelling is completed. Whilst every stage of construction should in itself be monitored, the total construction costs of a dwelling will have far more impact than the range of ups and downs within the individual stages. A compromise in this method of measurement is to use the units suggested in Figure 5.3 for each stage of external works but assess the number of dwellings as a measure to set against the external works totals. If all work listed under external works is 90% complete on a contract of 80 dwellings, the measure for external works would be 72 dwellings as indicated in Figure 5.2.

The house construction can then follow with each stage being measured in number of dwellings. If a housing contract involves the construction of vastly differing sizes of houses, flats, bungalows, etc., then a further alternative to counting numbers of dwellings can be used by taking the measure as the square metre of floor area in each building as defined by the local authority rateable area of the dwelling. Larger buildings will thereby contribute a larger measure which will go some way towards reducing the crudity of simply counting all dwellings as equal.

No matter what unit of measurement is used, all trades or stages of construction related to brickwork – foundation and superstructure, bricklaying, blocklaying and mortar supply – should at some point be measured in numbers of bricks laid (or in square metres of brickwork reduced to 1 B thick where this is the local custom). This can be carried out as a side note or comment where not embodied in the normal trade cost format. Blocks can be converted to bricks unless the allocation of labour hours between laying bricks and laying blocks is truly reliable. This conversion cannot be done on the basis of one square metre of block = one square metre of bricks as the work content in these two items is substantially different. However, if one accepts that it takes about twice as long to lay a brick wall as it does to lay the same thickness of blockwork, then blocks can be converted to equivalent bricks at the rate of $1\,m^2$ of 100 mm blockwork = 30 bricks. It follows that 150 mm blockwork converts at 45 bricks/m^2 and 200 mm at 60 bricks/m^2. Of course this conversion is only approximate but then the concept of trade costs is in itself a rough yardstick with no attempt at differentiating between individual items within a trade.

There is always a danger of output and cost statistics being quoted without any qualification, preamble or method statement. For this reason it is essential that all measurements reflect the quantity recoverable by the contractor prior to any allowances for waste, as for instance in excavation quantities where the measure must be the theoretical size of the hole dug with no allowance made in the measure for either bulking or overdig. All quantities must be physically measured and not based on delivery tickets as in the use of ready-mixed concrete. Work involved in making good or snagging must not be measured twice simply because it has been done again. Where the measure is being made of a risk item however, such as planking and strutting or formwork to bases, then the measurement should only reflect the quantity actually carried out with a side note of the risk factor, i.e. the percentage actually constructed of the possible total.

Labour hours

Each trade is to include chargehands, tradesmen, tradesmen's labourers, and apprentices appropriate to that trade. Hours are to be the actual hours worked exclusive of travel time, bus driving time, inclement weather time, overtime premium, greasing time or allocation sheet time. Apprentices' hours should be adjusted to take account of their lower rate of pay e.g. 40 hours of an apprentice on 90% of tradesmen's rate should be charged against the appropriate trade as 36 man hours. Hours paid in lieu of bonus as in a job and finish situation should be paid for as bonus and not recorded as man hours worked. Shift, daily and weekly paid labour are to be converted to approximate hours worked and included if involved in work covered by the headings of the trade cost.

Works staff up to and including the level of chargehands, gangers and slashers, usually in charge of one gang of men and machines are to be included but foremen and walking gangers, usually in charge of a number of gangs are to be excluded on the basis of being supervisory staff. Where foremen and walking gangers are paid on the workmen's payroll they should be listed under site overheads against a heading of Foremen on Workmen's Payroll.

Hired plant drivers and labour-only sub-contractors must also be included regardless of whether they are paid by the hour, shift or on measured rates. Where labour-only gangs from different sub-contractors are being used these should each be kept separate on the trade cost as should labour-only men from direct employees even if they are carrying out similar work. Where such men work together in the same gang then attempt should be made to apportion the measure of work done.

Totals of all hours set against the trade cost headings must be balanced with the total labour on the workmen's payroll adjusted for apprentices, labour on loan to or from the site, shift, daily and weekly paid labour, works staff, hired drivers and labour-only sub-contractors as illustrated in Figures 5.1 and 5.2.

Output – hours per unit

The output is to be calculated for each trade where a meaningful figure would result. As indicated in Figures 5.1 and 5.2 this is usually shown as hours per unit but can be expressed in whatever form is appropriate to the trade, e.g. laying bricks 45 No./h instead of the more cumbersome 0.022 h/No. or 22.22 h/1000 but fix rod reinforcement 25 h/tonne rather than 0.04 tonne/h. The output per man hour for bulk earthmoving and drainage included in Figures 5.1 and 5.2 have not been shown as these are substantially plant orientated items. For housing contracts costed in stages of construction rather than trades the output will be in hours per dwelling or h/m^2 of dwellings as indicated in Figure 5.2.

The outputs for site overheads work is best expressed as a percentage on the total number of hours of all productive work, for example, plant maintenance to date 283 h, total productive work 5721 h. Therefore the percentage hours to date on productive work for plant maintenance = 4.9 as indicated in Figure 5.1.

Labour cost

The actual cost to the company of all expenditure shown on the workmen's payroll including the cost of travel time, bus driving time, inclement weather time, overtime premiums, greasing time and allocation sheet time and the cost of plus rates, bonus, fares, subsistence, tool money, sick pay, public holiday pay, annual holiday stamps and the employer's contribution to the Graduated National Insurance (GNI) scheme. Such payroll items as the employee's contribution to GNI, union dues and PAYE taxes are not costs to the company but are deducted from the workman's gross earnings on his behalf. They must therefore be ignored in any cost calculations as they do not directly influence the cost to the company.

Redundancy pay shown on the workmen's payroll may create sudden surges of expenditure especially towards the completion of a contract and does not in any way represent the cost of that redundancy to the company, as a redundancy rebate (41% as of May 1982) is repaid to the company by the Government. Many contractors create a central account for paying redundancies and reclaiming government rebates, charging the residual costs to sites on a per capita or a percentage of payroll basis, either as part of the general head office costs or as a special *in-company redundancy fund*. If this is so then it is better to exclude redundancy payments from the trade cost. However, if such payments are not centralised they should be shown, *less* the appropriate government refund, as a separate heading under *site overheads work*.

The cost of apprentices should be charged against the work that they are doing even if the company operates a central apprentice funding system similar to the redundancy fund previously described. Although a company receives grants from the Construction Industry Training Board for apprentices, these are better offset against CITB levies paid by the company. This levy is not strictly a payroll item and on directly employed labour can generally be ignored for weekly costing purposes. However, on labour-only sub-contractors the levy is sufficiently high (2% as of May 1982) to warrant inclusion. The payroll cost to the company of shift, daily and weekly labour plus the payroll cost of works staff as defined under *labour hours* must be charged against the appropriate headings.

Cost of hired plant drivers will probably have to be assessed, as the rates quoted by plant hire companies for attended plant are usually for the plant, inclusive of driver, with no breakdown of labour and plant elements. This can be requested from the hirer but a site assessment is usually sufficiently accurate for costing purposes. Labour-only sub-contractor costs must be taken as the gross payment less any discounts but before deduction of any retention, PAYE taxes and CITB levy.

As with the labour hours the above costs must be balanced with the total expenditure on the workmen's payroll, adjusted if appropriate for redundancy payments, labour on loan to or from the site, shift, daily and weekly paid labour, works staff, hired drivers and labour-only sub-contractors as illustrated in Figure 5.1 and 5.2.

Plant cost

This cost is the actual charge of all company plant and plant hired from outside hirers and all associated consumable costs. Heavy plant such as cranes, excavators or compressors may be charged to sites under a different system from light plant such as barrows, drills, pumps or concrete skips. That is heavy plant may be charged by the hour whereas light plant may be charged by the week or month. It is therefore useful to separate these two distinct types of plant when balancing total plant costs.

Plant consumables should include all fuels, oils and greases and all spares, tyres and parts that are not included in the hire rates to the site. The majority of plant items can be allocated in total to an appropriate trade heading though some apportionment may be necessary where trades are sub-divided, for example, an excavator digging trenches and then lowering pipes into the trench as it proceeds if excavation and laying of pipes are recorded as separate trades.

Light plant, fuel and consumables will need to be apportioned to each trade and overhead heading. The simplest means of doing this is to relate these items to the heavy plant costs. In Figure 5.1 for instance, the week's total of company and hired heavy plant is £392 + £58 = £450, whereas light plant and consumables charges amount to £89 + £20 + £210 = £319 i.e. 71% on heavy plant costs. This is perhaps an oversimplification but is easy to operate and is usually sufficiently accurate for trade costs. Alternatively, light plant can be allocated item by item against different trades and fuel and consumables charged against each type of plant on the basis of either estimated or recorded consumption.

Cranes are generally better dealt with as an overhead item within material handling unless they have been brought on to a site as part of a production team to carry out a specific task such as handling precast concrete units in a precasting yard, pile driving or erecting components to timber framed houses, but not for the general craneage of a variety of

Brief Description of Work

Steel framed L-shaped single-storey factory. 60 m × 45 m × 8 m overall. Existing roads and surface drainage. Office and toilet facilities inside factory mainly of removable partitions. 12 varying machine bases 2.5 m deep mostly complex and all containing holding-down bolts

Period of Construction

12 months commencing 5 April, 1982

Industrial Relations

Go slow for two days by joiners early in contract

Weather

Exceptionally wet start to contract. Two weeks extension to contract granted

Quality

Client's COW was ex-bricklayer and demanded an exceptionally high quality of facing bricks

Abnormal Work

Concrete to one base pour failed test and was broken out. Costs included in appropriate trades – concrete £225, reinforcement £140, formwork £290

Excavation

Generally clay. 10% in made-up ground in NE corner of site. General excavation by JCB 806. Drains by JCB 3. No planking and strutting. 20 m^2 trench sheets required at drainage outfall 3 m deep. No pumping except at drainage outfall. All backfill with excavated material except broken brick hardcore to floor slab

Pipelaying

Generally 150 mm diameter Hepseal 1 m to 1.75 m deep

Manholes

675 mm diameter PC rings average 1.5 m deep handled by JCB 806

Services

Electricity, water and telephone, no gas

Roads
Existing concrete roads. Some breaking out for new drains and services

Concrete
Ready mixed. Foundations placed direct from delivery vehicle. Superstructure only 15% of total pumped by hired pumps. 90% floor area power floated average three passes

Reinforcement
Purchased cut and bent. Average diameter 15 mm. 20% particularly complex in machine bases. Fixed by labour-only sub-contractor

Formwork
All timber shutters. 25% crane handled by 22 RB

Holding-Down Bolts
12 sets of four bolts 50 mm diameter, 1.5 m long in machine bases, 36 sets of four bolts 25 mm diameter, 1 m long in ring beams

Brickwork
10% semi-engineering in foundations remainder cavity wall of red facings with recessed joint in black mortar and internal skin of 150 mm fair faced blockwork. Negligible internal partitions in 100 mm plastered blockwork

Material Handling
All materials unloaded in stores area. Fork lift truck used for transport. Small usage of 22 RB for material handling

Site Facilities
Facilities for six office staff, three works staff. No offices required for client or his representatives

Attend Sub-Contractors
Scaffolding required for cladding contractor

Figure 5.7 Method statement – ANY factory

reinforcement, formwork concrete and other materials. As with labour hours and labour costs the above plant costs must be balanced with the total site expenditure on plant as illustrated in Figure 5.1 and 5.2.

Total cost

The total cost shows the combined expenditure of labour and plant. It is not generally advisable to include the cost of materials in a trade cost, these are best dealt with as described in a separate procedure. However, if a complete cost of all elements of expenditure is required then the invoiced cost of materials plus allowance for deliveries not yet invoiced can be added at this stage preferably in the *to-date* records. The *this week* cost would then be the to-date figures *less* the previous week's to-date figures. This helps to avoid the confusion from week to week of deliveries and invoices not arriving on the same day.

Cost – pounds per unit

The cost per unit is to be calculated for each trade. This is to be expressed in £/unit and is to be the total cost divided by the measure for that trade, e.g. place concrete – measure $50\,m^3$ total cost £375, cost per unit £7.50/m^3 indicated in Figure 5.1.

For site overheads work the cost is best expressed as a percentage on the total cost of all productive work, e.g. plant maintenance to-date £1244, total cost of productive work £39 597, therefore percentage cost on total productive work to-date for plant maintenance = £3.1%. As Figure 5.1 indicates, this is different from the percentage calculated in hours.

To-date

The totals to-date are to be the accumulation of the weekly figures for measure, hours, labour, plant and total costs and are to be balanced against the appropriate to-date records of payroll, plant, costs, etc., in the same way as the weekly totals are balanced.

Method statement

No cost statistics are complete without some kind of method statement to explain the nature of the work on a particular contract and how that work was constructed. This does not need to be repeated each week and for site management is hardly necessary. However, for off-site records it is essential unless the users of the statistics are themselves intimately familiar with the contract and even then a simple method statement as illustrated in Figure 5.7 is useful as an *aide-mémoire*.

Chapter 6
Unit costing

The basis of unit costing is the computation of expenditure either in cash or in hours of an operation per single unit of measurement.

Example 6A Unit cost in man hours per unit

If six men take four hours to hand-dig an interceptor pit of size 3.5 m × 2.0 m × 1.0 m deep, i.e. 7 m³, then the unit cost in hours of digging the pit is

$$\frac{6 \times 4}{7} = \underline{3.4 \, h/m^3}$$

Example 6B Nett or basic unit cost

Assuming a basic labour rate of £1.62½/h the unit cost in cash would be

$$\frac{6 \times 4 \times 162\tfrac{1}{2}p}{7} = \underline{£5.57/m^3 \text{ nett}}$$

This unit cost is nett because it has been cashed out at the *basic* labour rate and does not include overheads of any kind.

A gross, or total payroll unit cost would be obtained by calculating first the gross cost of a man hour. This is achieved by dividing the cash total of the week's wages sheets by the number of hours worked. For instance, if the total of all costs to the company on the payroll was £1000 and the number of hours worked was 250 then the gross cost of an average man hour for that week is

$$\frac{£1000}{250 \, h} = \underline{£4.00/h}$$

Example 6C Gross or total payroll unit cost

The gross unit cost of excavating the interceptor pit during that week would therefore be

$$\frac{6 \times 4 \times £4.00}{7} = \underline{£13.71/m^3 \text{ gross}}$$

42 Unit costing

Figure 6.1 shows a sample sheet from a unit cost statement, the hours being the man hours spent on each operation day by day.

Operations involving the use of plant can be shown in a different colour to labour and the unit cost calculated in man hours per unit, plus plant hours per unit. In the illustration below plant has been shown encircled.

The conversion of the hours per operation into cash, either nett or gross, makes it possible to add together the labour and plant elements of an operation, whereas, when unit costs are presented in hours, this addition is only possible if plant hours are converted into the equivalent man hours, e.g. 1 No. 22RB hour = 3.75 man hours. Thus the operation of 'excavate basement by 22RB' illustrated in Figure 6.1 could be converted as shown in Figure 6.2.

W/E 21/2/82
WEEK NO 3 SHEET NO. 1

M	T	W	T	F	S	S	TOTAL HOURS	MEASURE	DESCRIPTION	UNIT COST
10	15		10	20			55	12 m³	Excavate bases by hand	4.6 h/m³
18 (9)	18 (9)	18 (9)	10 (4)				64 (31)	425 m³	Excavate basement by 22 RB (plant)	6.7 m³/h 13.7 m³/h
			8		15		23	3 m³	Concrete column bases	7.6 h/m³
9					9		9	2 m³	Blinding	4.5 h/m³
	7	18	21				46	7 m²	Fix shutters to columns	6.6 h/m²
18	9		6				33	6 m²	Make beam shutters	5.5 h/m²
				18	15		33	28 m²	Strip base shutters	1.2 h/m²
				9	8		17	8 m²	Make wall shutters	2.1 h/m²
18	18	18	18	18	15		105	45 m²	Brickwork to manholes 1B thick	2.3 h/m²

Figure 6.1 Unit cost statement in hours per unit

A typical unit cost form showing cash results is illustrated in Figure 6.3.

Unit costing provides immediate feedback of output data for use in future tenders; however, the regularity required for site cost control, usually weekly, is far too frequent for the contractor's estimators to revise their knowledge of outputs. Conditions on site, holidays, accidents, location, sickness, weather, etc., may vary a unit cost considerably from week to week, and only long-term information can be of use in the contractor's output library.

Where the prime purpose of the unit cost is to provide data for future estimating purposes or to record only a limited number of main items in the contract, then a record card for each major item can be set up keeping a running total of all measure and labour hours and/or labour cost appropriate to that item. Where plant plays a direct role in the production of the work this too can be included on the record card either being shown as a different colour or itemised separately as illustrated in Figure 6.4. In the example plant rates are shown inclusive of all fuels, oils and consumables and labour rates are shown inclusive of all payroll *on-costs* such as Graduated National Insurance, holiday stamps, etc.

Outputs can be expressed as man hours per unit or where the item is plant intensive the output can be expressed in machine hours per unit or more meaningfully as units per machine hour. Cumulative totals brought down can also be shown in a different colour or underlined as illustrated in Figure 6.4. Continuation sheets can be numbered as 15/A, 15/B, etc.

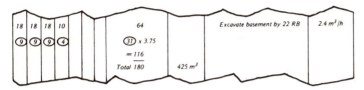

Figure 6.2 Unit cost statement with plant converted to equivalent man hours

M	T	W	T	F	S	S	TOTAL HOURS	@ 162½p = £ COST	MEASURE	DESCRIPTION	UNIT COST
10	15		10	20			55	89.38	12 m³	Excavate bases by hand	£7.45/m³

Figure 6.3 Nett or basic unit cost statement

For site monitoring purposes such unit cost record cards can be summarised each week as suggested in Figure 6.5 listing whatever elements of the cost or output are considered relevant to the particular site. Cumulative data can be included; however it is preferable not to produce long lists of items that have not changed for some weeks, therefore the summary should be restricted to only those jobs that have had work carried out during the week being costed. Unit costing, although more straightforward to operate than standard costing, has the following disadvantages.

(1) The system does not highlight the major losses, and without some means of comparison, either with previous weeks or with a library of average outputs, the problem of spotting the losing items is left to the more obvious operations where outputs are well known. The introduction of the metric system into the industry increases this problem, as years of well-learnt outputs have to be unlearnt for new metric units.

(2) The overall profit or loss of the contract is not known for the period of the unit cost. By addition the overall expenditure could be obtained but this would not indicate whether that expenditure were in excess of the contract's earnings for the period or otherwise.

(3) Although overheads can be included in the rate for calculation of a gross unit cost, there is no way of controlling the overheads as a

CONTRACT _____ UNIT COST RECORD CARD

DESCRIPTION OF ITEM __Excavate basement by 22 RB__ NO. 15

UNIT VALUE £1 – 50

Week ending	Labour				Plant				Measure in m³	Output m³/22RB h	Unit cost £/m³	Comments
	Hours	£ rate	£ cost	Details	Hours	£ rate	£ cost	£ total cost				
21/2/82	64	4·00	256·00	22 RB	31	10·00	310·00	566·00	425	13·7	1·33	
28/2/82	108	4·25	459·00	"	43	10·00	430·00	889·00	692	16·1	1·28	
	172		715·00		74		740·00	1455·00	1117	15·1	1·30	
7/3/82	84	4·10	344·40		38	10·00	380·00	724·40	550	14·5	1·32	
	256		1059·40		112		1120·00	2179·40	1667	14·9	1·31	
CARRIED FORWARD												

Figure 6.4 Unit cost record card

CONTRACT _____ WEEKLY SUMMARY OF UNIT COSTS W/E _7 March 82_

Card no	Description of items	Measure		Output		£ unit cost		£ unit value	Comments
		This week	To-date	This week	To-date	This week	To-date		
15	Excavate basement by 22 RB	550	1667	14.5	14.9	1.32	1.31	1.50	Output in m^3/ 22 RB h
16/A	Excavate bases by hand	10	45	2.1	2.4	7.50	8.30	8.00	Output in h/m^3
18	Concrete column bases	14	50	1.8	1.4	6.28	5.10	7.50	Output in h/m^3
20	Fix shutters to column	52	210	1.3	1.7	4.90	6.45	8.00	Output in h/m^2

Figure 6.5 Weekly summary of unit costs

						THIS WEEK				TO DATE		
M	T	W	T	F	S	S	TOTAL HOURS	MEASURE	DESCRIPTION	UNIT COST h/m^2	UNIT COST h/m^2	MEASURE m^2	TOTAL HOURS
7	18	21					46	$7 m^2$	Fix shutters to column	6.6	—	—	—
18	9	6					33	$6 m^2$	Make beam shutters	5.5	4.2	110	460
			18	15			33	$28 m^2$	Strip base shutters	1.2	1.2	212	253
			9	8			17	$8 m^2$	Make wall shutters	2.1	2.2	28	61

Figure 6.6 Unit cost statement showing cumulative figures

separate problem. For example, £13.71 per cubic metre gross calculated in Example 6C may be a good output for excavating in interceptor pit under particular site conditions, but perhaps the rate could be better if, for instance, fewer men were being paid lodging allowance. Without further investigation this question cannot be answered in a unit costing system.

(4) On the purely mathematical side there are two points against unit costing: first, the vast amount of divisional calculation required, division being extremely prone to human error; and second, the fact that calculations are not automatically totalled for balancing as a check against errors. Standard cost statements can, however, be easily balanced with the wage sheets, plant books, etc., to check that total expenditure is correct.

Use of unit costing as a control tool

Three basic methods are available for controlling costs by a unit cost system.

(1) From a basic unit cost statement as shown in Figures 6.1–6.3 site management can compare unit costs from week to week, watching for changes and studying further the items that suddenly show an increased unit cost. The statement can be extended to show the previous week's unit cost against each operation, which makes this comparison less difficult.

(2) By comparing the week's unit cost for an operation with the to-date unit cost for that operation, management can see whether the week's production is above or below the site's normal output. The reason for a sudden change can be researched and the appropriate action

TO DATE		
UNIT COST	MEASURE	TOTAL HOURS
1.8 h/m²	100 m²	180

Figure 6.7 Previous week's 'to-date' unit cost for brickwork

taken either to return production to normal or to record the interference causing the change and possibly recover the additional costs from the client. Figure 6.6 illustrates a weekly and cumulative unit cost statement. It can be seen that a comparison of the week's unit cost and the to-date unit cost quickly indicates which operations require attention. An even more effective comparison is that between this week's unit costs and the to-date unit costs of the previous week, i.e. before the to-date figures have been affected by this week's results.

48 Unit costing

Example 6D Comparison of this week's unit cost with previous to-date unit cost

The previous to-date records for an item of brickwork to retaining wall are as in Figure 6.7.

If this week's results are as in Figure 6.8, then the new to-date figures will be as in Figure 6.9.

			THIS WEEK		
S	S	TOTAL HOURS	MEASURE	DESCRIPTION	UNIT COST
		200	80 m²	Brickwork to retaining wall	2.5 h/m²

Figure 6.8 This week's unit cost for brickwork

TO DATE		
UNIT COST	MEASURE	TOTAL HOURS
2.1 h/m²	180 m²	380

Figure 6.9 New 'to-date' unit cost for brickwork

W/E 21/2/82   SHEET NO. 1

WEEK NO. 3

M	T	W	T	F	S	S	TOTAL HOURS	MEASURE	DESCRIPTION	UNIT COST	UNIT STANDARD
10	15		10	20			55	12 m³	Exc. bases by hand	4.6 h/m³	4.0
18 ③	18 ③	18 ③	10 ④				64 ㉛	425 m³	Exc. basement by 22 RB (plant)	6.7 m³/h (13.7 m³/h)	7.5 (15.0)
			8		15		23	3 m³	Concrete column bases	7.6 h/m³	6.0
9							9	2 m³	Blinding	4.5 h/m³	7.0
	7	18	21				46	7 m²	Fix shutters to cols.	6.6 h/m²	7.0
18	9		6				33	6 m²	Make beam shutters	5.5 h/m²	4.5
				18	15		33	28 m²	Strip base shutters	1.2 h/m²	1.0
				9	8		17	8 m²	Make wall shutters	2.1 h/m²	2.2
18	18	18	18	18	15		105	45 m²	Brickwork to manhole. 1B thick	2.3 h/m²	2.5

Figure 6.10 Unit cost statement with Standard Outputs included

The drop in output from the previous total of $1.8\,h/m^2$ to this week's $2.5\,h/m^2$ is more pronounced and more likely to arouse action that the drop within the week from an average of $2.1\,h/m^2$ to $2.5\,h/m^2$.

(3) By setting a target or standard output as being the norm for each operation and comparing the unit cost with this standard, the operations showing a loss on standard can be picked out and studied further to find the reasons for their losses. Figure 6.10 illustrates a unit cost statement with standards included.

Chapter 7
Standard costing

Standard costing is the calculation of expenditure, either in cash or in hours of an operation, and the comparison of that expenditure with a known standard or value for that amount of work or overhead.

Example 7A   Cost comparison in man hours

If six men take four hours to hand dig an interceptor pit of size 3.5 m × 2.0 m × 1.0 m deep, i.e. 7 m^3 then the cost of digging that pit is

6 men × 4 h = 24 man hours cost

If the contractor's standard value for carrying out this kind of work under similar conditions is 3.5 h/m^3, then the total standard value for this pit is

7 m^3 × 3.5 h/m^3 = 24.5 man hours value

Comparing the cost with the standard value

Value	24.5 man hours
Cost	24.0 man hours
Gain on operation	0.5 man hours

As with unit costing these results can be expressed either in nett costs, i.e. at basic rates, or in gross costs, i.e. including all payroll costs.

Example 7B   Nett or basic cost comparison

Nett cash value 7 m^3 × 3.5 h/m^3 × 162.5 p/h	£39.81
Nett cash cost 6 men × 4 h × 162.5 p/h	£39.00
Gain on operation (nett)	£00.81

Standard costing 51

Example 7C   Gross or total payroll cost comparison

Gross cash value 7 m³ × 3.5 h/m³ × £4.00/h	£98.00
Gross cash cost 6 men × 4 h × £4.00/h	£96.00
Gain on operation (gross)	£2.00

Figure 7.1 shows a sample sheet from a standard cost statement, the hours being the man hours spent on each operation day by day.

Site management is left in no doubt as to which operations are below standard, as they are readily listed in the loss column of the standard cost statement. For example

Excavation to bases losing 7 h
Excavation to basement losing 7 h labour

W/E 21/2/82											SHEET NO. 1		
WEEK NO. 3													
M	T	W	T	F	S	S	TOTAL HOURS	MEASURE	@ h	STANDARD VALUE, h	DESCRIPTION	HOURS GAIN	HOURS LOSS
10	15		10	20			55	12 m³	4	48	Exc. bases by hand	—	7
18 (9)	18 (9)	18 (9)	10 (4)				64 (31)	425 m³	2/15 (1/15)	57 (28)	Exc. basement by 22RB (plant)	— ⊖	7 (3)
			8		15		23	3 m³	8	24	Concrete column bases	1	—
9							9	2 m³	4	8	Blinding	—	1
	7	18	21				46	7 m²	8	56	Fix shutters to columns	10	—
18	9		6				33	6 m²	5	30	Make beam shutters	—	3
				18	15		33	28 m²	1.5	42	Strip base shutters	9	—
			9	8			17	8 m²	3	24	Make wall shutters	7	—
18	18	18	18	18	15		105	45 m²	2½	112	Brickwork to m/h. 1B thick	7	—
TOTAL LABOUR							385			401		34	18
TOTAL PLANT							(31)			(28)		⊖	(3)

Figure 7.1 Standard cost statement in man hours

Excavation to basement losing 3 h plant
Blinding losing 1 h
Make beam shutters losing 3 h

In addition to the individual performance of each operation by straightforward addition, it is possible to calculate from the standard cost what the contract's performance as a whole has been for the week being studied.

In the example, gains of 34 h and losses of 18 h have been made on labour, giving an overall contract gain of 16 labour hours for the week with an overall loss of three plant hours.

When the standard cost is calculated in man hours, it is possible only to include those overheads that can be calculated in man hours, e.g. travelling time, non-productive overtime, chainman, erection of temporary buildings, etc. For these a standard value can be calculated and a gain or loss shown. However, there are also numerous overheads that cannot be expressed in hours (except by converting pounds into equivalent hours), e.g. lodging allowance, holidays with pay, stamps, etc.

The means of calculating standards for overheads are dealt with in Chapter 10. Figure 7.2 shows assumed standards illustrating a standard cost statement in nett cash for both measured work and overheads.

W/E 21/2/82 WEEK NO. 3											SHEET NO. 1		
M	T	W	T	F	S	S	TOTAL HOURS	= £ COST	MEASURE	@ £	STANDARD VALUE £	DESCRIPTION	GAIN £	LOSS £
10	15		10	20			55	Labourers @ £1.62½ = 89.38	12 m³	6.75	81.00	Exc. bases by hand	—	8.38
18 (9)	18 (9)	18 (9)	10 (4)				64 (31)	= 104.00 @ 10.00/h = 310.00	425 m³	0.60	255.00	Exc. basement by 22 RB (plant)	—	159.00
		18	27	27	9		81	Carpenters @ £1.90 = 153.90			100.00	Erect. temp. offices	—	53.90
from wages sheets								25.00			40.00	Tea woman	15.00	—
from wages sheets								24.00			16.00	Non-prod. Overtime	—	8.00
from wages sheets								205.50			150.00	Site staff	—	55.50

Figure 7.2 Standard cost statement in nett or basic cost

For practical purposes it is not necessary to quote cash figures to the nearest penny; gains or losses of less than £1.00 are relatively insignificant and can normally be ignored. Paperwork can thus be reduced if all extensions both of cost and of value are completed to the nearest pound. A further saving in computation can be made by averaging the cost per hour of tradesmen and labourers, in the ratio applicable to the contract. Within

M	T	W	T	F	S	S	TOTAL HOURS	@ £1.75 = £ COST	MEASURE	@ £	STANDARD VALUE, £	DESCRIPTION	GAIN £	LOSS £
10	15		10	20			55	96	12 m³	6.75	81	Exc. bases by hand	–	15

Figure 7.3 Use of average cost rate per hour

CONTRACT _____ W/E 31/1/82
SHEET NO. 1 WEEK NO. 28

COST CONTROL STATEMENT

DESCRIPTION OF WORK	VALUE £	COST £	GAIN £	LOSS £
Piling (Labour)				
Drive 300 mm x 300 mm piles	600	450	150	–
Drive 450 mm x 450 mm piles	150	160	–	10
	750	610	150	10
Piling (Plant)				
Drive 300 mm x 300 mm piles	450	400	50	–
Drive 450 mm x 450 mm piles	100	110	–	10
	550	510	50	10
Excavator (Labour)				
Excavate vaults	180	295	–	115
Excavate bases	60	250	–	190
Excavate trenches	75	70	5	–
	315	615	5	305
Excavator (Plant)				
Excavate vaults 22 RB	100	150	–	50
Excavate bases 22 RB	40	75	–	35
Excavate trenches JCB 3C	90	80	10	–
	230	305	10	85
Concretor (Labour)				
Blinding	400	300	100	–
Concrete bases	350	275	75	–
Drain bed and surround	100	75	25	–
Fence post holes	40	35	5	–
Columns	80	70	10	–
Beams	20	15	5	–
Ground floor slab	90	60	30	–
	1080	830	250	–
Formwork (Labour)				
Bases	50	60	–	10
Bases (Credit)	250	–	250	–
Columns	50	40	10	–
Beams	55	35	20	–
Ground floor slab stop ends	75	65	10	–
	480	200	290	10
Formwork (Sub-let)				
Columns	150	75	75	–
Beams	210	125	85	–
Suspended slab	90	60	30	–
	450	260	190	–
Reinforcement (Sub-let)				
Cut and bend reinforcement	525	600	–	75
Fix bars	455	525	–	70
Lay fabric	75	50	25	–
	1055	1175	25	145
Bricklayer (Labour)				
Manholes	125	100	25	–
Scaffolding (Labour)				
Independent scaffolding	125	110	15	–
Scaffold ramps	20	20	–	–
	145	130	15	–

Figure 7.4 Trade summary of cost control statement

the limits of accuracy required this figure can then be rounded off to an easily calculable figure, e.g. £1.75 per hour, as illustrated in Figure 7.3.

It can be seen that this approximation causes a slight inaccuracy; but as the total expenditure on the standard cost sheet can at the end be balanced with the total expenditure on the wages sheets, etc., the amount of this inaccuracy can be readily seen and, if necessary, corrected. In practice the degree of control is rarely so fine that these approximations become critical.

CONTRACT		W/E 31/1/82		
SHEET NO. 2		WEEK NO. 28		
COST CONTROL STATEMENT				
DESCRIPTION OF WORK	VALUE £	COST £	GAIN £	LOSS £
Variable Overheads				
Non-productive overtime	160	340	—	180
Importation of labour	295	520	—	225
Inclement weather time	60	—	60	—
Holiday with pay stamps	170	150	20	—
Employer's Graduated National Insurance	190	170	20	—
National Increases	130	—	130	—
Plus rates and extras	125	160	—	35
	1130	1340	230	440
Non-Variable Overheads				
Unload materials	100	340	—	240
Tea and canteen	25	100	—	75
Chainman	30	30	—	—
Plant maintenance	50	300	—	250
Pumping	25	70	—	45
Clean public roads	20	125	—	105
Clean office	10	60	—	50
Attend on Clerk of Works	10	60	—	50
Transport on site	50	250	—	200
Temporary lighting	15	—	15	—
	335	1335	15	1015
Plant Overheads				
Concrete mixer	30	40	—	10
Mortar pan	15	15	—	—
Coaches	60	110	—	50
Transport on site	50	200	—	150
Lighting set-up	5	—	5	—
Pumps	10	25	—	15
Vibrators	15	20	—	5
Power float	10	5	5	—
Compressor	15	10	5	—
	210	425	15	230

Figure 7.5 Overheads summary of cost control statement

In gross cash costing the gross labour rate is calculated at regular intervals, preferably weekly to tie in with the wages sheets, and will therefore reflect any shift in the labourer-to-tradesmen ratio. It is possible to calculate such a gross labour rate for different trades by separating the trades on the wages sheets. This may be necessary where, for instance, a particular trade is being paid exceptionally high bonuses, thus causing say a 10% increase on the average rate for all other trades.

Standard cost statement

On many contracts the number of gangs and intermixing of items necessitates the rewriting of the numerous operations carried out each week under trade or section headings rather than gang headings – this provides an opportunity of reducing the cost statement to an easily readable form and thus encouraging site management to put it to use rather than to bed!

CONTRACT					W/E 31/1/82			
					WEEK NO. 28 OF 104			
	COST CONTROL SUMMARY SHEET							
	THIS WEEK				TO DATE			
	VALUE	COST	GAIN	LOSS	VALUE	COST	GAIN	LOSS
MEASURED WORK								
Piling (Labour)	750	610	150	10	10120	8700	1600	180
(Plant)	(550)	(510)	(50)	(10)	(550)	(510)	(50)	(10)
Excavator (Labour)	315	615	5	305	6030	5400	1100	470
(Plant)	(230)	(305)	(10)	(85)	(10100)	(9065)	(1300)	(265)
Concretor	1080	830	250	—	5360	4300	1200	140
Reinforcement	—	—	—	—	4300	3150	1850	700
Formwork	480	200	290	10	3400	3375	75	50
Bricklayer	125	100	25	—	3500	3250	250	—
Drainage	—	—	—	—	1250	950	350	50
Scaffolder	145	130	15	—	1500	1475	25	—
Measured Bonus	—	—	—	—	—	—	—	—
Non-Productive Bonus	—	290	—	290	—	2400	—	2400
TOTAL MEASURED LABOUR	2895	2775	735	615	35460	33000	6450	3990
TOTAL MEASURED PLANT	(780)	(815)	(60)	(95)	(10650)	(9575)	(1350)	(275)
TOTAL MEASURED	3675	3590	795	710	46110	42575	7800	4265
OVERHEADS								
FIXED LABOUR PRELIMS.	335	1335	15	1015	8450	16235	615	8400
PLANT PRELIMS.	(210)	(425)	(15)	(230)	(3500)	(4930)	(145)	(1575)
ON COSTS	1130	1340	230	440	15375	14390	3790	2805
TOTAL OVERHEADS	1675	3100	260	1685	27325	35555	4550	12780
SUB-LET								
Reinforcement	1055	1175	25	145	2250	2525	275	550
Formwork	450	260	190	—	450	260	190	—
TOTAL SUB-LET	1505	1435	215	145	2700	2785	465	550
GRAND TOTAL	6855	8125	1270	2540	76135	80915	12815	17555

```
GAIN/LOSS   THIS WEEK    £1270   —   18½ %
GAIN/LOSS   TO DATE      £4780   —    6¼ %
```

Figure 7.6 Final summary of cost control statement

Figure 7.4 shows a typical standard cost statement rewritten under trade headings. Note the way information is kept to a bare minimum; this promotes action by encouraging the eye to run down the final column and thus observe the losses. For additional effect the losses can be shown in red, or the losses column can be tinted red and the gains column tinted green. Such visual aids, although amusing at first, are bound to attract attention, which is all-important when a particular site manager is reluctant to use modern management tools or no follow-up system is built into the standard costing scheme as suggested in Figure 16.1.

Overheads, both labour and plant, can be grouped together according to type and listed on a similar statement as illustrated in Figure 7.5. All statements can then be summarised and totals kept up to date of the various trades and overheads, as illustrated in Figure. 7.6.

Standard costing in civil engineering

Where the plant element of a contract is likely to be extensive, e.g. in motorway and other large earthmoving contracts, the method of splitting the occasional items of plant illustrated in Figure 7.4 becomes a little clumsy; two sets of figures continually have to be added together for each operation before a realistic comparison can be made. Labour and plant values and costs can be added together for such a statement, although provision will have to be made elsewhere for the following factors.

(1) In nett costing the total labour value is required for calculating the variable overheads. This can easily be abstracted from the standard cost statement when labour and plant are separated but will have to be noted separately if they are combined. This can be done in the margin or in an additional column. Alternatively, the variable overheads can be related to the labour *and plant* element of the contract. However, this is a somewhat false relationship and would show unrealistic results on a contract having a large plant element; for example, extensive earthmoving by scraper would increase standard value for holiday with pay, stamps, etc., because of increased plant standard value in measured work. Thus control of variable overheads would not be possible, as they would not be directly related to pure labour values.

(2) Balance of labour costs with wage sheets would not be directly possible. However, if the cost of plant on site is recorded separately, a balance can be made between total costs in the standard cost statement and wages sheets plus plant costs plus cost of labour-only sub-contractors.

Alternatively additional columns can be drawn across the form to produce totals for labour, plant and combined L + P cost figures.

Exercises

(1) Discuss the costing systems known to you and decide whether they are standard costing or unit costing or a mixture of the two systems.
(2) State which of the two systems, standard costing or unit costing, you would prefer to use as a management tool and give details why.

Chapter 8

Monitoring expenditure

The three elements of expenditure described in Chapter 1 as requiring regular control are

(1) labour costs
(2) plant charges
(3) site overheads

A breakdown of expenditure for these three elements is required before either a unit costing or standard costing system can commence; i.e. hours of plant and labour require allocation against individual operations of work. There are a number of ways of obtaining this allocation.

Labour costs

Task sheets

Daily task sheets may be used for each man or gang of men. Figure 8.1 illustrates a task sheet designed for issue to each man as he clocks on in the morning.

As he starts or finishes each operation, the worker notes down the approximate time, say to the nearest 15 or 30 min; alternatively he notes down the duration of each operation. Should his programme be altered for some reason, he can easily add additional items to the programmed list as in the example: the 're-erection of the signboard accidentally knocked over by the contractor's plant'.

Where a gang of men are all carrying out the same operations with only occasional variation by members of the gang, then this sheet can be satisfactorily used for that gang with any variations noted on the sheet.

For example, in an excavation gang of six men two of the gang stay behind in a base excavation to bottom up (level out the bottom of the base) while the others move to their next job of work. This can be shown as in Figure 8.2.

If durations only are shown, the format can be that of Figure 8.3.

The larger the gang and the more diverse the operation, the more difficult such a simple task sheet becomes to control. A pour of concrete, for instance, may have a large number of attendance items required

58 *Monitoring expenditure*

before, during and after the main function of placing the concrete: for example, preparation of construction joints; travelling crane from last concrete pour; preparation of barrow runs for area out of crane's reach; preparation of vibrators; pouring and vibrating concrete; tamping concrete; clearing up surplus concrete and barrow runs; trowelling; rubbing up; removing wet sand used for curing.

ANY FIRM & SON LTD.		CONTRACT	
DAILY TASK SHEET			
PROGRAMME FOR DAY		START	FINISH
Remove door from Superintendent's office	D/W	9.00	9.30
Rehung door when safe installed	D/W	3.00	4.00
Fix skirtings to Rooms 23 and 24		8.00 / 9.30	9.00 / 11.00
Hang doors A13 and A14		11.00 / 1.00	12.30 / 3.00
Fix pipe casing in NW corner of Room 11		–	–
Re-erect signboard knocked down by J.C.B.		4.00	5.00
NAME: D. Peters		DATE: 31/1/82	

Figure 8.1 Daily task sheet showing times recorded for one man

PROGRAMME FOR DAY	START	FINISH
Excavate base 3C	8.00	10.30 (4) / 11.30 (2)
Excavate drain trench	10.30 (4) / 11.30 (2)	12.00

PROGRAMME FOR DAY	NO. OF MEN	DURATION
Excavate base 3C	4	2½
	2	3½
Excavate drain trench	4	1½
	2	½

Figure 8.2 Method of recording times of number of men on daily task sheet

Figure 8.3 Daily task sheet showing durations

For this reason, and to obtain greater accuracy, it is better to show the duration of *each* man on *each* operation.

Figure 8.4 illustrates a gang task sheet as written up by the planning engineer or similar person, and based on the site's overall programmes and resources.

Short-term planning is duscussed in Chapter 27, however, it is necessary to point out the uselessness of producing daily task sheets based solely on programme requirements without first discussing the work with the foreman concerned. His intimate knowledge of available resources, shortcuts, probable absenteeism, individual's capabilities, etc., can turn a random list of operations into an enthusiastically attempted programme of work.

Labour costs 59

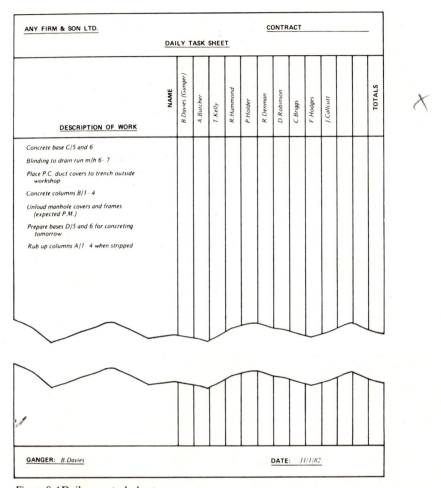

Figure 8.4 Daily gang task sheet

Where a job card system (as discussed in Chapter 26) is in operation for incentive purposes, times can be entered on the job card either against a total job or allocated against the various tasks within that job. For example, shutter walls – stop ends, surface fixings, height lath and joint key.

Ganger's allocation sheets

Where site planning or team co-operation have not yet reached the level necessary to produce daily task sheets for a contract, and therefore individual operations cannot be pre-listed, an attempt must be made to classify work under the required descriptions after that work has been carried out. This can be done by relying on the ganger to write down the descriptions of what his men have been doing.

Figure 8.5 illustrates a ganger's daily allocation sheet similar to the daily task sheet but completely written out by the ganger or foreman. As with

60 Monitoring expenditure

the gang task sheets, a column is allowed for the time spent by each man on each operation. The reduction of columns to a single gang column, although reducing the size of the form, would undoubtedly either create confusion by the ganger attempting to reconcile, for instance, 1 man @ 3 h, 2 men @ 6 h, 2 men @ 4 h, and 1 man @ 2 h, concreting the basement slab, or, more likely, lead to approximations that might end up by the ganger roughly spreading the man hours among the operations, disregarding all limits of accuracy.

Expenditure Data for Unit Costing and Standard Costing

ANY FIRM & SON LTD. DAILY TIME ALLOCATION SHEET												CONTRACT	
DESCRIPTION OF WORK	NAME → P.Foot (Ganger)	A.Purchese	R.Fulton	L.Winter	V.Ireland	B.Smith	J.McIlroy	M.Davidson	L.Dorrington				TOTALS
Concrete basement slab	3	6	6		4	4	2						
Unload bricks	1						1	1	1				
Concrete columns	2				4	4	2						
Concrete over drains	1	1	1		1	1							
Moving manhole covers and frames from stores to site	1	2	2										
Tea boy							4						
Bush hammer floor slab for Grano paving	1							8	8				

(column: Absent — under L.Winter)

GANGER: P.Foot   DATE: 31/1/82

Figure 8.5 Daily allocation sheet

When written out as suggested by site men, the legibility and phraseology of these sheets can, not surprisingly, be something of a problem, though not without its lighter side. Comments of 'lifting off Irish Jays' and 'burning out Polish Irene', are perhaps understandable, although the recently seen entry of 'sissy engineers' when the chainboy was sick and a labourer came to the engineer's assistance makes one wonder whether the degree of illiteracy is not intentional!

By using (a) standard descriptions or (b) coded descriptions, it is possible to alleviate this problem to some extent, though at the same time creating others.

Standard descriptions are in a way an extension of the daily task sheet; but instead of listing only those operations programmed for the day, the sheet lists every possible operation that could be carried out on the contract. For reasons of pure length the lists have to be split into trades or some other division, and even then can amount to some half-dozen sheets of possibilities for a single trade. With carpenters, for example, make, fix and strip, and every different type of shutter are listed separately. With excavators dig, plank and strut, trim and backfill are listed separately. This, of course is true only of a large contract, but then a small contract hardly requires the use of standard descriptions. As with the daily task sheets, space must be left at the end for the ganger to write in unlisted work. Figure 8.6 illustrates a standard description allocation sheet for the steelfixer trade on a particular contract.

Coded descriptions remove the problem of incomprehensible allocation sheets by allocating a code to every operation; thus the ganger or foreman only has to write down the code and will not be tempted to ramble on into illegibility.

A possible code might start with the identification of the trade

E.	— Excavation	R.	— Reinforcement
C.	— Concrete	D.	— Drains
F.	— Formwork	B.	— Brickwork
			etc.

The next two letters of the code would show the location

ab.	— Abutment	dr.	— Drain
bs.	— Base	sl.	— Slab
bt.	— Basement	wl.	— Wall
bm.	— Beam	rf.	— Roof
cl.	— Column		etc.

For example, C.cl. means concreting to columns; C.bt.rf. means concreting to basement roof.

For the majority of items this kind of code is sufficient; and provided that details are placed in a prominent position on the wall of the ganger's or foreman's office, it is quite possible to operate such a scheme. However, a lot depends on the temperament and education of the ganger or foreman, and it is suggested that an engineer or timekeeper assists in the filling-in of such an allocation sheet.

An alternative suggestion is a numerical code

(1) Excavation
(2) Concrete
(3) Formwork
(4) Reinforcement
(5) Drainage
(6) Brickwork
etc.

62 *Monitoring expenditure*

.01 Oversite
.02 Reduce levels
.03 Basement
.04 Bases
.05 Footings

with further figures giving dimensional or other further details.

For example, 1.02 means excavate to reduce levels; 4.04 means reinforcement to bases.

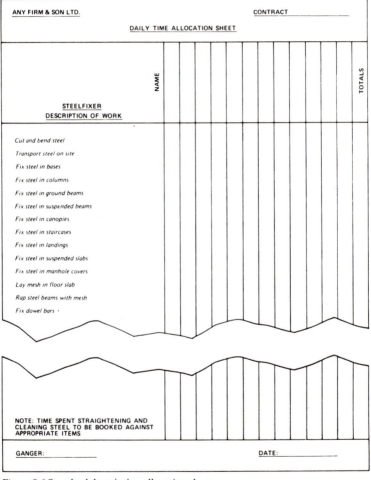

Figure 8.6 Standard description allocation sheet

A numerical code as suggested above is usually not capable of being used at foreman level, allocation being made in the normal long-hand way and coding being added at office level later for subsequent handling by computer.

In an attempt to operate the ganger's allocation system speedily and accurately, a ganger is frequently given some additional incentive, e.g. an additional ¼ hour's pay per day or an additional few pence on his hourly rate. It is important that the conditions under which the ganger or foreman receives this payment be fully known and enforced. For example, payment only made if allocation sheet is on agent's desk by 10.00 am on day following work being allocated. No sheet – no pay, late sheet – half pay.

Where bonus schemes are in operation the tendency of the ganger may be to book hours against non-bonusable operations; for this reason an attempt must be made to bonus all work so that the ganger will realise that nothing can be gained from this practice.

Perusal of the allocation by the ganger's or foreman's superior will discourage inaccurate or sloppy allocation; and as a check on the accuracy of the system, a spot cost on an individual gang during the course of a day will prove sufficient provided that the foreman is not aware that the check is being carried out.

Personal observation

Personal observation of the distribution of men is made by one of the following three basic methods.

(1) Instead of the ganger or foreman filling in an allocation form, the responsibility is taken over by a timekeeper, engineer or surveyor, and it is then his duty to patrol the site at frequent intervals in order to note down any changes in labour and plant activities. This system is necessary where site personnel are illiterate or are reluctant to become involved in paperwork. It also alleviates the problem of unrecognisable descriptions, and enables the timekeeper, engineer or surveyor to become familiar with the workings of the site. It is clear that on even a medium-sized contract such personal observation could become a full-time operation, and on an extensive development or civil engineering contract could require such a large staff that the method would be uneconomical.

(2) Staff requirements can be reduced by requiring the timekeeper, engineer or surveyor to patrol the site less frequently, say once during the morning and once during the afternoon. The frequency is now too long to be able to record sufficiently accurate movements of labour. Some operations may be completely carried out between observations and would therefore be missed. However, the personal observations can be supplemented by verbal information from the ganger or foreman. A five-minute chat between say, the timekeeper and the ganger is a lot less expensive than an hourly trip round the site by the timekeeper.

(3) On some civil engineering sites allocation of resources may be unnecessary, as a gang may be working on the same operation for a number of weeks or even months. Observations are then necessary only for method study or for incentive bonus payments. Costing can therefore be carried out directly from the timebook.

Plant charges

Except for special studies the majority of small plant items on a site do not need to be allocated hourly. Little is to be gained by allocating, for instance, a poker vibrator to each individual pour of concrete or a pump to the various items of excavation. These can be charged to the contract on a weekly basis and costed as *plant overheads*. Large plant, on the other hand, can be allocated on an hourly basis in the same manner as labour, as illustrated in Figure 8.7.

ANY FIRM & SON LTD.							CONTRACT			
DAILY TIME ALLOCATION SHEET										
NAME	P. Holder (Dr)	L. Hennessy (B/M)	22 RB							TOTALS
Excavate basement	4	4	4							
Excavate lift base	5	5	5							

Figure 8.7 Labour and plant allocation sheet

Mixers, tower cranes, etc., can be similarly allocated, though each contract and each item of plant poses the question 'Is the information to be gained from such allocation worth the effort involved?' It is not necessary during a regular costing system to know how long a mixer is working for a certain pour of concrete, nor what a tower crane is lifting during each hour of the day. This kind of study is best left as a method study exercise and not incorporated in the costing system. It is suggested therefore that allocation of plant be generally confined to excavation and piling equipment only.

The rates used for calculating the cost of plant, whether allocated or weekly, must be the true cost to the contract. If the plant is hired, then the hire charge applies; if the plant is owned by the contractor, then the contractor's internal hire rate applies. It is not normally necessary to approximate between these two sets of charges unless, for instance, two or more identical machines are being used on the site, one hired and one contractor owned. The use of two separate rates may cause unfair weighting against one operation in favour of another. For example, two of the contractor's own excavators are digging bases but have lost approximately two weeks of programme. Duct excavation is due to start and cannot be delayed until one of the excavators becomes free in two weeks' time. The contractor has no further machine available; therefore a machine is hired for two weeks and put to work excavating the ducts.

It would be a true cost to charge the higher rate against the duct excavation but a fairer result might be obtained by averaging out the excavator rates during the two-week period that the hired machine is on site. A similar situation exists where two different contracts require the same machine so the contractor has to hire an additional machine at a higher rate than his own. All other things being equal these two contracts would inevitably show a different unit cost for identical work carried out by these machines.

Where this extra cost of hiring creates major discrepancies it may be preferable to use company hire rates for costing productive work even if hired machines are used and to show the 'additional cost of hiring' as a plant overhead item, there being probably no allowance for this made at the time of tender so no budget or value can be set against the cost.

Another item of expenditure which can be included under the heading of plant overheads is the cost of fuel, oils, spares, etc. This is preferable to adding an estimated cost of fuels, etc., to the plant hire charges rate, because

(1) the plant rate then becomes partly fact and partly estimated, and
(2) any losses of fuel costs would be concealed by the plant costs.

A difficulty arises when plant is standing owing to lack of work, to breakdown or to lack of driver. Should the means of hiring or the contractor's internal hiring policy necessitate a charge to the contract during standing or broken-down time, then this charge is best shown as a plant overhead and possibly labour overhead of *standing time*. It is unrealistic to levy such charges against any particular operation that happened to be under way at that time or, worse still, happened to be showing a good output at that time. Note should be taken here of any differentials applicable to plant charges when the plant is not working or of any minimum working time agreements between the contractor and the plant hire-company or department.

Site overheads

Site overheads fall into three basic categories

(1) wage sheet overheads
(2) allocated labour overheads
(3) plant overheads

Wage sheet overheads

Nett costing or basic rate level of costing

For use in standard cost statements the wage sheet overheads can readily be abstracted from the wages sheets: for example, total cost for week of lodging allowance, sickness payments, holiday stamps, etc.

Flat rate overtime working must be allocated in the same manner as normal working hours, but the premium time (excess overtime or non-productive overtime) requires to be kept separate and calculated as an individual cost item.

Wage sheet overheads cannot be included in nett unit costing except as an addendum list at the end of the unit cost statement.

Gross costing or total payroll level of costing

Where the unit or standard cost is to be produced 'gross' as described in Chapters 6 and 7, then wage sheet overheads become part of the average gross rate and are not studied individually.

Allocated labour overheads

Nett costing or basic rate level of costing

Whatever system is used for allocating labour costs hour by hour, certain non-measurable items will be disclosed which are best dealt with as an overhead cost. These items include erection of site office, unload materials, lay temporary roads, clean public roads, etc. In unit costing these items can only be listed as an addendum to the unit cost statement; however, in standard costing it is possible to incorporate such items into the general standard cost statement under the heading of overheads.

Gross costing or total payroll level of costing

In both unit costing and standard costing these items can be listed as an addendum to the gross cost statement. Alternatively, when the average hourly rate is calculated the wage sheets can be divided by the total of the productive work hours only, i.e. total hours less allocated labour overheads; the gross rate will then be slightly higher and will include these overheads.

Plant overheads

Nett costing

Reference has already been made to suggested items of plant overheads (page 64). Such items as vibrators, fuel and oils, mixers, dumpers, etc., can be abstracted from the site plant book, records of plant on site or invoices without need for allocation. Plant standing or broken down can be recorded in the same way that the plant working is allocated. It is not usual to record the cost of vibrators, dumpers, etc., standing as a separate item, as these are already an overhead and are probably charged anyway on a weekly basis.

In unit costing these items, too, can only be listed as an addendum to the unit cost statement; however, in standard costing it is possible to list such items under a heading of plant overheads.

Site overheads

Gross costing

Plant overheads are usually dealt with as described under nett costing, though as an extension of the gross labour rate an addition can be made to cover overhead plant charges, i.e. the calculation of the gross labour rate

$$\text{g.l. rate} = \frac{\text{wage sheet total} + \text{overhead plant charges}}{\text{number of productive hours worked}}$$

The result can be quite a high average gross rate out of all proportion to the basic labour rate but nevertheless including all overhead costs, both labour and plant.

Exercises

(1) Design a system of collecting labour and plant allocation for use on each of the following sites. What instructions would you issue for ensuring correct working of the system?

£620 000	Block of flats
£120 000	6 No. detached houses
£12 000 000	10 km motorway
£3 000 000	Sewage works
£30 000	Car factory maintenance
£12 000 000	Reinforced concrete dam

(2) Bulk earthmoving requires special attention due to the high plant costs and therefore critical need for high outputs. Draw up a form for collecting the allocation of earthmoving plant by personal observations (load checking), bearing in mind that the different types of excavation – topsoil, suitable clay, unsuitable ripped mudstone, etc. – and different haul lengths and condition of haul route require to be recorded by the checker if any worthwhile feed-back of information is to be gained.

(3) Design a job card suitable for recording detailed allocation of tasks within that job and for indicating which men have carried out the job.

Chapter 9

Standards for measurable work

The previous chapter has shown the source of the various elements of cost for use in both unit costing and standard costing.

However, the basis of a standard costing system is the comparison of these costs with some yardstick or standard, and the question immediately arises: 'With what can the costs be compared?'

Source of standards

The standards used are likely to be derived from one or more of the following

(1) library of historical outputs
(2) work study standards
(3) estimator's pricing
(4) textbook pricing

Library of historical outputs

Under factory conditions it is possible to record consistency of outputs which will never be achieved on a construction site. However, this lack of consistency is not a reason for ignoring historical records but more a reason for conserving as much data as possible under all types of conditions on all types of sites, the relevant conditions being recorded with the output as indicated in Figure 9.1.

Extreme differences of conditions are worthy of separate recording; for instance, it is pointless to list together excavate bases in sandstone and excavate bases in granite. The average of these two items would be a meaningless figure. The cards or forms containing similar types of items, as in this example, can, however, be filed next to one another, and a handy record of outputs for similar items under differing conditions will thus be preserved.

It is important to indicate any special points about a contract which may have affected the average output recorded – for example, exceptional

OPERATION		J.C.B. EXCAVATION TO BASES AND PITS IN CONSOLIDATED FILL								
DATE		CONTRACT	DESCRIPTION OF CONTRACT	SPECIAL REMARKS	QUANTITY	LAB. HOURS	J.C.B. HOURS	HOURS	HOURS	OUTPUT
FROM	TO									
Jan 79	Mar 79	Office Block	R.C. frame £1 000 000	Industrial area. Approx. 25% broken brick in fill Aver. base size 6 m² x 3 m deep	450 m³	135	40			11¼ m³/h
Feb 80	June 80	Factory	Steel frame £500 000	Unsuitable fill from motorway. 75% clay Aver. base size 4 m² x 3 m deep	380 m³	120	35			11 m³/h
July 81	Dec 81	Bowling Alley and Skating Rink	R.C. frame £1 500 000	Industrial area Approx. 10% ash in fill. Aver. base size 9 m² x 4 m deep	800 m³	170	65			12½ m³/h

Figure 9.1 Record of company outputs achieved

weather conditions, abnormal requirements of architect or engineer, use of explosives, errors not possible to separate from output, use of new methods, etc. Even the name of the foreman in charge may be found to have a consistent effect on the output. It is important, too, that the calculation of the output be over as long a period as possible in order to avoid the recording of brief abnormalities. The ideal period for recording output data is no less than from start to finish of the operation being studied. This reveals the only true average output: at the middle of the operation everything is in full swing; at the beginning the operation is having teething troubles; at the end of the operation cleaning up and finishing off is being carried out.

Work study standards

Where work study engineers are employed or a work study data bank is available, as for example, the local authority data bank known as LAMSAC or the data produced by the British Management Data Foundation (29 St James's Street, London SW1A 1HB), an ideal situation exists for comparing actual costs with a scientifically calculated standard. It must, however, be emphasised that work study standards derived from one contract must be used on other contracts with extreme caution, methods must be fully defined in all records and a consistent approach given to relaxation allowances, supervisory time, travelling time, labourers in attendance and unoccupied time which occurs during team work when a man or machine is forced to be idle even though the man : machine ratio is at the optimum.

For use in standard costing either the 'allowed time' or 'standard time' can be used, provided the percentage difference is clearly understood, standard time being based on a motivated rate of working (BS 3138:1979 Clause 43032) and allowed time (BS 3138:1979 Clause 54033) being an easing of that standard usually for incentive scheme or cost control purposes. On a poorly organised site, a site with little incentive or low team morale, few actual outputs are likely to reach the work study standard, thus making allowed time a more realistic yardstick for control.

Estimator's pricing

Where no adequate record of outputs is available either from past costing records or from work study, an alternative basis must be found for comparison with costs, the most likely alternative being the data used in pricing the contract tender.

Consideration must first be given to whether the standard cost based on estimator's rates is to be run

(1) as a *pure standard cost*, i.e. by simply using the estimator's standards as a guide and varying the standards according to conditions as found on the site, or

(2) as a strict *profit and loss account* adhering rigidly to the prices and outputs used in the tender, varying only on the issue of a variation order by the client.

For the majority of operations these two methods are identical; but without the historical data necessary to review standards scientifically the frequent varying of standards suggested in (1) above can be abused and can lead the costing system into fantasies that are not respected by site management. It is preferable therefore, when estimator's standards are used, to adhere strictly to known figures whatever the circumstances and whatever the changes on site, as suggested in (2) above.

In breaking down the estimator's rates, care must be taken to return all adjustments to their appropriate places ready for use in the costing system under the correct heading.

Various adjustments may have been made to the estimator's original calculations; these generally fall under the following headings.

Addition of preliminaries to the rates

The initial cost of setting up offices, mixers, site roads, etc., may have been added to the excavation items in order to recover the cost early in the contract; before excavation rates are used as costing standards, the allowance for these items must be deducted, to leave only the excavation element of the rates.

Internal company adjustments

The pricing of a contract may have been adjusted to allow for using written-off plant; the plant standard must therefore be nil or whatever nominal amount has been left in the pricing. If other plant has to be used, e.g. hired plant, then a loss will correctly show up against the standard for the operation using that plant.

Feel of the market

Because of a full order book, because of the need to acquire work to give continuity, because of goodwill, or because of the feeling that other contractors are fighting for a contract or are not interested in a contract, a firm may well adjust a series of rates or the final figure of their quotation or tender. If a true standard cost system is to be operated, this adjustment must not affect the setting of the standards; however, if a profit and loss standard cost system is to be operated, it is preferable to make this adjustment where appropriate, thus observing the strict rule of no variations on known figures. It can be argued, however, that such an adjustment as this cuts into or enhances the planned profit margin and should therefore not affect anticipated outputs. This is, of course, true but increased or decreased output have nevertheless got to be achieved to make up for the adjusted profit margin.

Client's effect

An adjustment may be made within a quotation or tender because of the anticipated ease or difficulty of working with a certain client, architect or consultant. Such an adjustment must also be made to the costing standards in anticipation of higher or lower outputs during construction.

Error

Errors in pricing usually have to be accepted by the contractor and in a profit and loss standard cost must be carried through to the standard. In a pure standard cost, however, the error can be corrected and the standard can be made realistic. With the introduction of the metric system the old bogey of pricing yard super items with a foot super price has, of course, disappeared, but care must still be taken to see that pounds have not become pence or thicknesses of concrete, timber, brickwork, etc., misread.

Late information

Although last-minute quotations from sub-contractors and suppliers do not usually affect the labour, plant and overheads section of the pricing, a late labour-only sub-contractor may well submit a price in time for an adjustment to be made on the tender figure. Such an adjustment must be carried through to the costing standards for sub-let work.

Deviation from normal

For example, excavation rates may as a matter of course include an allowance for pumping. If this is removed or increased on the tender figure because the site is exceptionally dry or exceptionally wet, then this adjustment must also be made to the standard for pumping.

Weighting of money

For financial reasons a contractor may reduce the rates for certain items, transferring the money to other items in the Bill of Quantities or quotation. This money must be returned to its rightful place before standards are calculated. Even where the standard cost is of the profit and loss account type, this deviation from the billed prices back to estimator's prices is desirable, as curious gains and losses will obviously result from the use of weighted rates or outputs. Any increases or decreases on billed quantities of weighted items can be studied outside or as an addendum to the normal standard cost system, e.g. abstracted from the monthly valuations.

Breakdown of estimator's rates

Methods of estimating vary tremendously throughout the industry; from rates worked out on the back of a cigarette packet to rates calculated entirely by computer. Rates may be based on a hunch or on work study;

Estimator's pricing 73

but however they have been arrived at, a breakdown must be carried out to discover the standards that will subsequently be used in the standard costing system. Two elements require extracting from the bill or quotation rate – labour and plant – either gross or nett, depending on the proposed standard cost.

Because a standard cost is an attempt to compare like with like, the planned profit margin allowed in the estimate must be removed if standards are being calculated in cash. The same applies to the allowance for head office overheads, unless these overheads are being included on the cost side of the standard cost, e.g. added into the calculation for the gross labour rate. The method of estimating will dictate where and when this deduction takes place.

BILL ITEM 23/R

Hardcore fill to lift-shaft in 250 mm layers

			£/m³
MATERIAL			
Approved broken brick	£5.00/m³		
Add 20% compaction	£1.00/m³		
	£6.00/m³		
+ 5% profit	30p		
	£6.30		6.30
LABOUR			
Spread, level and roll			
2 h/m³ @ £3.50/h gross			7.00
PLANT			
Vibratory hand roller			
10 m³/h @ £1.50/h			
+ 5% profit 7½p			
	£1.57½		£0.16
			£13.46/m³

Figure 9.2 Estimator's build-up of rate

Figure 9.2 illustrates an estimator's pricing of an item for hardcore fill to a lift shaft and is based on the gross labour rate shown in Example 10.1.

Breakdown for gross standard costing for this item is as shown in Figure 9.3.

Gross labour standard is 2 h × £3.18, i.e. labour rate before addition of profit and head office (HO) charges calculated in Example 10.1.

Gross plant is 1/10 × £1.50. No overheads have been added to plant in this estimate.

	£
Other elements are the addition of	
Material	6.30
Profit on labour and HO charges – 2 h @ 32p	0.64
Profit on plant – 1/10 × 7.5p	0.01
	£6.95

74 *Standards for measurable work*

The final column is provided as a balancing column to check mathematical errors.

Breakdown for nett standard costing for the same item is as shown in Figure 9.4.

Nett labour standard is 2 h × £1.62.5p, i.e. basic labour rate.
Nett plant is ¹⁄₁₀ × £1.50.
Other elements are the addition of

	£
Material	6.30
Overheads, profit on labour and HO charges 2 h @ £1.87.5p	3.75
Profit on plant – ¹⁄₁₀ × 1p	0.01
	£10.06

BILL ITEM	BILL RATE, £	GROSS LABOUR STANDARD, £	GROSS PLANT STANDARD, £	OTHER ELEMENTS INCL. ALL PROFIT AND H.O. CHARGES, £
23/R	13.46	6.36	0.15	6.95

Figure 9.3 Breakdown of estimator's price for gross costing

BILL ITEM	BILL RATE, £	NETT LABOUR STANDARD, £	NETT PLANT STANDARD, £	OTHER ELEMENTS INCL. ALL PROFIT OVERHEADS AND H.O. CHARGES, £
23/R	13.46	3.25	0.15	10.06

Figure 9.4 Breakdown of estimator's price for nett costing

Breakdown by hours instead of cash for the same item is as shown in Figure 9.5.

By breaking down each item to be costed in this way it is possible to compile a complete list of labour and plant standards for every measurable operation envisaged during the course of the contract.

BILL ITEM	LABOUR STANDARD h/unit	PLANT STANDARD units/h	TYPE OF PLANT
23/R	2 h/m³	10 m³/h	Vib. roller £1.50/h

Figure 9.5 Breakdown of estimator's price for costing in hours

Textbook standards

Where neither an adequate library of standards nor estimator's calculations are available, e.g. in spec. building, package deal or where a lump sum contract has been negotiated, then the only remaining source of standards is from one of the many textbooks on estimating backed by an individual's records and knowledge of outputs.

This system of setting standards, although approximate, does provide a degree of control over the site expenditure in that costs are compared with a constant and variations from that constant are easily seen and can be further studied.

Feed-back of standards

To be able to build up an adequate library of output standards, a costing system must be geared to recording the relevant historical data. The ideal period for recording output data is no less than from start to finish of the operation being studied. The costing system must therefore be arranged to collect measure of work done and totals of labour and plant expended for each operation being costed. This can be carried out at the completion of each contract, but the more preparatory work that has already been done the better. It is no easy task to filter through months, even years, of site costs in an attempt to abstract measurements and expenditure for like items that may be described quite differently by different personnel and yet could have been easily collected together under the same heading as work was being carried out.

CONTRACT									
FEED-BACK RECORD SHEET									
STANDARD VALUE	£0.60		£0.80		£0.90				
ITEM	Lay and joint 100 mm dia. concrete pipes		ditto 150 mm		ditto 300 mm				
WEEK NO.	m	h	m	h	m	h			
15	30	10							
16	45	13	10	6					
	20	8	25	15					
17	35	13							
	28	10			10	7			
18			20	13	10	6			
19	14	6							

Figure 9.6 Feed-back record sheet

76 Standards for measurable work

To ensure that such collections of data are prepared as the work proceeds, the feed-back records can be incorporated into the costing system by using them as the collecting sheet for similar work by different gangs. This *feed-back* sheet can also be an information sheet containing the relevant labour and plant standards, as illustrated in Figure 9.6.

It is possible to abstract task sheets or allocation sheets, etc., directly on to the feed-back record sheet, as shown in Figure 9.7. This procedure,

CONTRACT									
FEED-BACK RECORD SHEET									
STANDARD VALUE	£1.00		£1.25		£2.00		£2.50		
ITEM	50 mm paving slabs in large areas		Ditto to decorative pattern		50 mm gravel paths hand rolled between paving slabs		250 mm x 50 mm p.c. edging incl. backing with concrete		
WEEK NO.	m²	h	m²	h	m²	h	m²	h	
28 M		16							
T		4		12					
W				16					
T		8						8	
F				8				8	
S				8					
SUB-TOTAL	66	28	70	44			10	16	
29 M				16					
T				16					
W				16					
T				16					
F						16			
S						8			
SUB-TOTAL			101	64	30	24			
30 M						8		8	
T						8		8	
W		16							
T		16							
F		16							
S						8			
SUB-TOTAL	89	48			26	24	11	16	

Figure 9.7 Feed-back and allocation collection sheet

however, is paper-consuming (continuation sheets may have to be written out each few weeks) and requires daily searching through an ever-increasing number of sheets to find each item of entry. It is therefore recommended only for small contracts and would be unsuitable for systems incorporating an incentive scheme, as the weekly sub-totals would have to be abstracted on to a separate form for bonus calculations. It is advisable before setting up a system of feed-back to decide on which items are to be recorded, what they are to include and what is to be ignored, otherwise inconsistencies will occur from contract to contract. Feed-back of operations using plant should be further separated by types of plant. For example, trench excavation by JCB 3 and JCB 806 recorded together will in total give a meaningless output. At the end of a contract the feed-back records shown in Figures 9.6 and 9.7 can be totalled and, with the addition of general comments applicable to the contract as a whole, the master records completed as illustrated in Figure 9.1.

If it is required to obtain outputs that reflect the possible rather than the actual, then inefficiencies during the contract must be kept separate. This requires the system of labour and plant allocations (Chapter 8) to be sufficiently detailed to enable lost time to be isolated. Alternatively, outputs can be calculated weekly and those outside a permitted deviation from the current mean excluded from feed-back records as being questionable.

Where work study standards are used for production control then not only must output be recorded but also overall performances (BS 3138:1979 Clause 51036) of various trades and type of contract, in order that the data can be used for overall planning and estimating. It is optimistic to assume that 100 overall performance will always be maintained even under incentive conditions.

Learning curves

The output per man hour of one gang of men working for 100 days is not the same as the output of 100 gangs working for one day. Although such extremes are an unlikely option on a construction site this statement helps to illustrate the thinking behind a learning curve which arises not so much because the workmen have not done that job before, but because they gain experience of doing it in the situation and within the gang of men with whom they are currently working. It is important therefore that some perspective is given to any output data recorded by stating also the quantities involved.

Group studies

(1) Within the definitions given in BS 3138:1979 *Glossary of Terms Used in Work Study*, compare the various systems of adding allowances to basic times which are in use by companies and advisory services in the construction industry to produce standard and allowed times.
(2) Discuss the various methods of estimating known to you and the problems arising when these estimates are later used as a basis for standard

costing. Decide on the method most suitable for both build-up and breakdown, bearing in mind that not every tender is successful.

Exercises

(1) Assuming that no output library is available and that a factory contract has been placed on a price per square metre, without detailed estimating, use any available textbooks, etc., to research and calculate standard outputs in h/m^2, for use in a standard costing system for the following items of shuttering. Quantities have been roughly measured. Calculate, make, fix and strip separately.

300 m^2 ground beams
500 m^2 pile caps
100 m^2 edge of 200 mm floor slab
100 m^2 construction joint in 200 mm floor slab
350 m^2 walls
250 m^2 columns
300 m^2 suspended beams
100 m^2 suspended slabs
10 m^2 suspended landings

(2) Show how your approach to standards would differ when using estimator's prices as the basis of a true standard cost system from using the system as a profit and loss account.

(3) Break down the following five examples of bill pricing under the headings of labour and plant (a) nett, and (b) gross, where 10% of the addition for profit and overheads accounts for profit and head office overheads.

(i) *Estimated from first principles*

19 mm tongued and grooved boarding nailed to joists

	£
Material	
T and G boarding	5.00
Waste – 7.5%	0.38
Nails	0.20
	5.58
Profit – 5%	0.28
per m^2	£5.86

	£
Labour	
Unload and fix	
Carpenter 0.30 h @ £1.90	0.57
Labourer 0.15 h @ £1.62.5p	0.24
	0.81
Profit and overheads – 120%	0.97
Material	5.86
per m^2	£7.64

(ii) *Estimated from sub-let prices*

1 : 2 : 4 concrete in bases

	£
Material	34.00
+ 5% profit	1.70
	35.70
Labour and Plant	
Mix and place concrete sub-let	6.00
Add profit and overheads – 20%	1.20
	7.20
per m³	**£42.90**

(iii) *Estimated from labour constants*

215 mm brickwork in 1 : 3 cement mortar

Material   £

Bricks $\dfrac{£60.00}{1000} \times \dfrac{125}{1}$ (including waste)   7.50

Mortar 0.1 m³/m² (1B) @ £25.00 per m³   2.50

Total materials   10.00

+ 5% profit   0.50

                                                                                                                                         £10.50

Labour   £   £
Standard £60.00 per thousand   60.00
Add 10% for difficulty on this contract   6.00

                                                                                                          66.00

$£66.00 \times \dfrac{115}{1000}$ 7.59

Profit and overheads – 120%   9.11

                                                                                                                         16.70
Material   10.50

per m²   **£27.20**

(iv) *Estimated nett with profit and overheads added at end of Bill of Quantities*

BILL REF.	QUANTITY	LABOUR	PLANT	MATERIAL
16/A	250 m²	4.50	0.75	18.75
B	35 m²	4.75	0.75	18.75
		+ 120%	+ 5%	+ 5%

Figure 9.8

(v) *Estimated off the cuff*

Excavate interceptor pit £6.25 per m³

Chapter 10

Standards for site overheads

Although historical data on site overheads may prove useful in estimating such items as importation of labour, attraction money, sick pay, etc., in certain regions or for certain types of contract, the majority of overheads will be peculiar to each contract and will usually be calculated from first principles.

The amounts allowed in an estimate for overheads are therefore the most reliable means of calculating standards for overheads. In the absence of such an estimate, budgeted figures must used based on the same reasoning that an estimator would have used had he been pricing the contract in detail.

Site overheads can be divided into two distinct types: (a) fixed and (b) variable.

Fixed overheads

Few overheads are rigidly fixed during the full period of a contract but many remain more or less unchanged by the current labour strength at any particular time in the contract. These are termed fixed overheads, and include items of both labour and plant. Examples are

Site offices	Chainboy
Toilets	Attendance on clerk of works
Temporary roads	Clean public roads
Staff	Plant maintenance
Minibuses	Temporary services
Tea woman	Small tools

These items are frequently priced under the preliminaries section of the Bill of Quantities and are therefore often referred to as *preliminaries* in a standard costing system.

The estimated cost or budget for these overheads can be spread over the contract to obtain weekly standards. The following examples show how that budget can be spread.

(1) Weekly allowance, e.g. cleaner for offices.
   £25.00 per week allowed for duration of contract.
(2) Limited weekly allowance, e.g. chainboy.
   £30.00 per week for first 20 weeks of contract.
(3) Lump sum allowance, e.g. initial site survey.
   £750.00 allowed for survey.
(4) Measurable allowance, e.g. scaffolding.
   50p allowed for scaffolding per square metre of brickwork constructed.
(5) Random allowance, e.g. temporary roads.
   £1500 total allowed to be drawn on at random.
(6) Allowance related to value of work, e.g. minibuses.
   3% on nett labour value of work completed.
(7) Combination of above, e.g. temporary offices.
   £1250 erect offices,
   £500 dismantle offices,
   £25 per week hire for duration of contract.

Variable overheads

A number of overheads are geared directly to the number of men employed on site and vary as that number changes. These are termed variable overheads and include such items as

Annual holiday stamps	General insurances
Public holiday with pay	Sick pay
Inclement weather time (IWT)	Redundancy pay
Travelling time allowance	Graduated National Insurance
Fares	National increases
Lodging allowance	Training levy

Some of these items such as general insurances and training levy, may not be dealt with at site level and would therefore become part of the head office charges not included in the standard cost.

The pricing of these variable overheads is frequently calculated as an on-cost to the basic labour rates and the Bill of Quantity items priced gross, i.e. including the appropriate share of these overheads as in the pricing of *hardcore to lift-shaft* (Figure 9.2), which is based on the labour rate calculated in Example 10A. Variable overheads are therefore often referred to as *on-costs* in a standard costing system.

Example 10A   Build-up of hourly rate for labourers

Labourer's rate

Basis 45 hours worked/week; 52-week year *less* 6 weeks' holiday, 1 week sick, 2 weeks' wet, 3 weeks' absenteeism = 40 weeks/year

	£
Labourer's basic rate	1.62½
Plus rates (average)	0.02
Bonus (26p/h guaranteed minimum) say naturally paid	0.50

Non-productive overtime $\dfrac{6 \times \frac{1}{2} \times £1.62\frac{1}{2}}{45}$	0.11
Public holidays $\dfrac{(8\text{ days @ say 8 h}) @ £1.62\frac{1}{2}}{40\text{ weeks} \times 45\text{ h}}$	0.06
Sick pay $\dfrac{\text{average 1 week} \times 6\text{ days @ £4.49}}{40\text{ weeks} \times 45\text{ h}}$	0.01½
Inclement weather time $\dfrac{\text{average 2 weeks} \times 39\text{ h @ £1.62}\frac{1}{2}}{40\text{ weeks} \times 45\text{ h}}$	0.07
Travel time allowance average 16 km = $\dfrac{60\text{p}}{9\text{ h/day}}$	0.06½
Total subject to Graduated National Insurance	*2.46½*
Employer's GNI (within earnings limit of £27.00 to £200.00) 13.7% of £2.46½	0.34
Allowance for redundancy and training levy 2% of £2.46½	0.05
Fares allowance (no company transport provided) Average 16 km = $\dfrac{96\text{p}}{9\text{ h/day}}$	0.10½
Subsistence	Nil
Holiday stamps for annual holidays £9.00/45 h	0.20
Miscellaneous payments £1.00/45 h	0.02
	£3.18
Add profit and HO costs 10%	32
	£3.50

Gross labourer's rate £3.50/h

In Example 10A for every £1.62½ worth of labour work carried out the contract will earn an additional 50p to cover the cost of bonus and other additional sums to cover the other variable overheads. The 2p (plus rates) can be expressed as 1.23% of £1.62½, which provides an accurate relationship between the value of an individual overhead and the value of measured work completed for the current week or period being studied. The percentage of £1.62½ for all the variable overheads shown in Example 10A are as follows

Plus rates	1.23%
Bonus	30.77%
Non-productive overtime	6.77%
Public holidays	3.69%
Sick pay	0.92%
Inclement weather time	4.31%
Travel time	4.00%
Employer's Graduated National Insurance	20.92%
Redundancy and training levy	3.08%
Fares	6.46%
Subsistence	Nil%
Holiday stamps	12.31%
Miscellaneous payments	1.23%
Total variable overheads	95.69%

A check can be made to ensure mathematical accuracy as follows

£3.18 − £1.62½ = £1.55½ overheads
95.69% of £1.62½ = £1.55½ overheads

These figures differ for each contract but need calculating once only at the commencement of the costing system. Exmaple 10A shows the variable overheads for labourers; Example 10B illustrates a similar calculation for joiners.

Example 10B Build-up of hourly rate for joiners

Tradesmen's rate (joiners)

Basis 45 hours worked/week; 52-week year *less* 6 weeks' holiday, 1 week sick, 1 week wet, 2 weeks' absenteeism = 42 weeks/year

	£
Tradesmen's basic rate	1.90
Tool money 85p/45 h	0.02
Bonus (31p/h guaranteed minimum) say actually paid	0.70
Non productive overtime $\dfrac{6 \times \frac{1}{2} \times £1.90}{45}$	0.12½
Public holidays $\dfrac{(8\ \text{days}\ @\ \text{say}\ 8\ \text{h})\ @\ £1.90}{42\ \text{weeks} \times 45\ \text{h}}$	0.06½
Sick pay $\dfrac{\text{average}\ 1\ \text{week} \times 6\ \text{days}\ @\ £4.49}{42\ \text{weeks} \times 45\ \text{h}}$	0.01½
Inclement weather time $\dfrac{\text{average}\ 1\ \text{week} \times 39\ \text{h}\ @\ £1.90}{42\ \text{weeks} \times 45\ \text{h}}$	0.04
Travel time allowance average 30 km = $\dfrac{1.44}{9\ \text{h/day}}$	0.16
Total subject to Graduated National Insurance	3.02½
Employer's GNI (within earnings limit of £27.00 to £200.00) 13.7% of £3.02½	0.41½
Allowance for redundancy and training levy 2% of £3.02½	0.06
Fares allowance (no company transport provided)	
Average 30 km = $\dfrac{£1.89}{9\ \text{h/day}}$	0.21
Subsistence	Nil
Holiday stamps for annual holidays £9.00/45 h	0.20
Miscellaneous payments £1.00/45 h	0.02
	£3.93
Add profit and HO costs 10%	39
	£4.32

Gross joiner's rate £4.32/h

Expressed as a percentage, the tradesman's variable overheads are as shown in Figure 10.1. Based on a contract average of one labourer to one tradesman, the average percentages for all men on this contract are as shown in Figure 10.2. Thus if the total nett value of measured work for a week is £10 000 the contract will also have earned 6.68% × £10 000, i.e. £668 as the value for non-productive overtime. Similarly 3.55% × £10 000, i.e. £355 is the value to be put by for public holidays and 0.86% × £10 000, i.e. £86 for sick pay.

Basic rate	£1.90	100%
Tool money	0.02	1.05%
Bonus	0.70	36.84%
Non-productive overtime	0.12½	6.58%
Public holidays	0.06½	3.42%
Sick pay	0.01½	0.79%
Inclement weather	0.04	2.11%
Travel time allowance	0.16	8.42%
Employers GNI	0.41½	21.84%
Redundancy and training levy	0.06	3.16%
Fares	0.21	11.05%
Subsistence	Nil	Nil%
Holiday stamps	0.20	10.53%
Miscellaneous payments	0.02	1.05%
Total variable overheads	2.03	106.84%

Figure 10.1 Variable overheads for joiners expressed as a percentage on basic rate

Travelling time, fares and lodging allowance may be grouped together under a heading of Labour Importation. Holiday stamps, however, are best considered in isolation as being a fixed amount, they can easily expose a situation of too many men on site where the value of measured work cannot support the cost of the number of stamps each week.

Exercises

(1) Calculate gross rates for labourers and tradesmen based on current basics and express the variable overheads as average percentages of the average basic rate.
(2) State the differences in approach to setting standards or budgets for fixed and for variable-type overheads.

	Labourers	Tradesmen	Average
Basic rate	£1.62½	£1.90	£1.76½
Plus rates	1.23%	–	0.61%
Tool money	–	1.05%	0.53%
Bonus	30.77%	36.84%	33.80%
Non-productive overtime	6.77%	6.58%	6.68%
Public holidays	3.69%	3.42%	3.55%
Sick pay	0.92%	0.79%	0.86%
Inclement weather	4.31%	2.11%	3.21%
Travel time allowance	4.00%	8.42%	6.21%
Employers GNI	20.92%	21.84%	21.38%
Redundancy and training levy	3.08%	3.16%	3.12%
Fares	6.46%	11.05%	8.76%
Subsistence	–	–	–
Holiday stamps	12.31%	10.53%	11.42%
Miscellaneous payments	1.23%	1.05%	1.14%
Total variable overheads	95.69%	106.84%	101.27%

Figure 10.2 Variable overheads for average workforce expressed as a percentage on basic rate

Chapter 11
Standard cost example

The following example follows the basic principles of a standard cost from collection of site data (in this case by ganger's allocation sheet) through the bonus system and feed-back records to a weekly cost statement. The final summary sheet of trades and overheads shows also the *to-date* picture of the contract, so that trends in overheads, financial progress compared with programme and overall gains and losses can be observed. The previous week's cost summary sheet, i.e. for week six is shown in Figure 11.1 in order that to-date figures can be carried forward.

It will be noted that gains and losses are not cancelled out until the end of this final summary. This procedure is adopted in order not to hide items or trades that are losing money by offsetting those losses against other items that show a gain.

The example chosen illustrates a nett cost presented in cash. The advantages of costing in cash rather than hours are as follows.

(1) Labour-only sub-contractors are usually paid on a measure basis which is not directly related to the number of hours they have worked.
(2) If an item of plant different from or additional to the one anticipated is used, a complicated calculation has to be carried out to convert hours value of one machine into hours value of another or additional machine. In cash the value remains constant. Where standards are taken from a standards library, however, the alteration of method or plant is more simple, since reference only has to be made to the library for the revised standard.
(3) Cost of fuel oils and consumables can be added to the hire cost of plant rather than treating these plant on-costs as separate overhead items.
(4) Many overheads can only be expressed in cash.
(5) The impact made by the cost statement is psychologically more effective when shown in £ than in hours.

There are, however, two main disadvantages of cash records.

(1) Allocation of domestic labour and plant to various operations commences in hours; thus it requires conversion into cash at some point in the system if the final statement is to be in cash.

	FORM NO C/5
CONTRACT _____	W/E 21 March 82
	Wk No 6 of 75

COST CONTROL SUMMARY SHEET

	THIS WEEK				TO DATE			
	Value	Cost	Gain	Loss	Value	Cost	Gain	Loss
MEASURED WORK								
Excavator	374	609	11	246	410	656	11	257
Excavator (Plant)	(333)	(520)	(23)	(210)	(378)	(580)	(23)	(225)
Concretor	45	50	–	5	45	50	–	5
TOTAL MEASURED LABOUR	419	659	11	251	455	706	11	262
TOTAL MEASURED PLANT	(333)	(520)	(23)	(210)	(378)	(580)	(23)	(225)
TOTAL MEASURED	752	1179	34	461	833	1286	34	487
OVERHEADS								
Fixed labour prelims	245	404	–	159	670	984	41	355
Plant prelims	(36)	(36)	–	–	(72)	(72)	–	–
On Costs on £664 value	673	525	227	79	1140	870	392	122
TOTAL OVERHEADS	954	965	227	238	1882	1926	433	477
SUB-LET								
Excavator	243	238	5	–	261	255	6	–
TOTAL SUB-LET	243	238	5	–	261	255	6	–
GRAND TOTAL	1949	2382	266	699	2976	3467	473	964

Gain/Loss This week £ 433 = 22.2 %

Gain/Loss To date £ 491 = 16.5 %

Figure 11.1 Final summary of cost control statement for week 6

(2) Feed-back of output data required to be in hours rather than cash, otherwise continual and undesirable up-dating of cash standards is necessary. This means that if the final statement is in cash, then outputs must be converted back into hours for record purposes; however, the example avoids this additional work by collecting feed-back data before the hours are converted to cash, i.e. before the standard cost statement.

Bonus scheme

Chapter 25 discusses the various types of incentive scheme; however, the bookeeping requirements are almost identical regardless of the type of scheme and are indeed very similar to those of standard costing in as much as the expenditure for a task or series of tasks is compared with a target for that work. In the example shown in Figure 11.16 to 11.19 an hours saved system is used with savings paid out as bonus at the rate of £1.50 per man hour or in the case of the 22 RB target at the rate of £5.00 per machine hour saved.

In the example, not all men in the gangs are given an equal portion of the bonus earned, the amount depending on the number of hours worked by each man with the ganger and 22 RB driver receiving 1¼ shares per hour compared with the labourer's one share per hour.

The bonus targets are in man hours unless otherwise stated and are as follows

Excavate basement by 22 RB	15 m³/22 RB hour
Excavate bases	3½ h/m³
Excavate man holes and drains	4 h/m³
Excavate oversite	4½ h/m³
Blinding	5 h/m³
Concrete bases	3½ h/m³
Concrete drains	5 h/m³
Concrete slab	6 h/m³
Unload bricks	1½ h/thousand
Unload cement	½ h/tonne
Erect hut	18 h (assessed)
Chainboy	6 h (assessed)

Measurement of work done

Measurements for the week ending 28 March, 1982, i.e. week 7 are as follows

Excavate basement and cart away	1000 m³
Excavate bases and cart away	50 m³
Excavate manholes and drains and cart away	28 m³
Excavate oversite and cart away	10 m³
Blinding	7 m³
Concrete bases	13 m³

Concrete drains	4 m³
Concrete slab	6 m³
Unload bricks	14 000 No.
Unload cement	9 tonne

All excavations are carted away by a labour-only sub-contractor but after bulking the quantity paid for on the lorry is 1300 m³.

Standard cost example

Figures 11.2–11.7 show allocation sheets as written out by the ganger J Green in charge of the general labouring gang. Absentees are included on the sheets but are shown as 'A'. Figures 11.8–11.14 show similar sheets completed by C Harris, the 22 RB driver, allocating for himself, his banksman and his machine. The men and machines shown on these sheets must be checked daily against the time-books or clocking system and discrepancies queried.

As a second check that man hours are correct, comparison should be made with the wage sheet totals illustrated in Figure 11.15 which also shows the summary of wage sheet items required for cost control of variable overheads. In practice this summary can be a stumbling block to the whole system, as wages may well be calculated by remote computer or by clerks who have other duties, thus not providing the summary until late in the week. However, the cost statement for productive work can be prepared and corrective action taken where indicated prior to the addition of the wage sheet items. Overheads are by their nature a more long-term problem and do not therefore have the same urgency as the calculation of bonuses and preparation of the cost statement for productive work.

The work done and hours taken by J Green's gang are transferred from the *daily time allocation sheets*, Figures 11.2 to 11.7 and entered each day on the weekly sheet shown in Figure 11.16, the first entry being for the 43 h spent by the gang on Monday in excavating bases. The second entry combines both dig drains 9 h and dig manhole 4 h together under the description of excavate drains. The total of 79 h allocated on Monday's daily time allocation sheet, Figure 11.2, must agree with the total hours for Monday on the weekly sheet Figure 11.16.

As the week progresses, so the quantities of work done and the appropriate bonus targets can be entered. After the week ending, the last day's allocation can be included together with any remaining quantities and targets. Having checked that the total of 413.5 man hours for J Green's gang balances with the payroll hours for the gang, the total target can be calculated by multiplying each of the quantities of work done by the appropriate target. The total of these target hours, 518, can then be compared with the actual man hours to produce the number of hours saved i.e. 104.5 as illustrated in Figure 11.17. In this instance each hour saved is paid at a bonus rate of £1.50 which produces a total bonus for the gang of £156.75 to be shared in whatever ratio has been agreed.

A similar process is then adopted for Harris's gang, the only difference being that the calculations for bonus shown in Figures 11.18 and 11.19 are

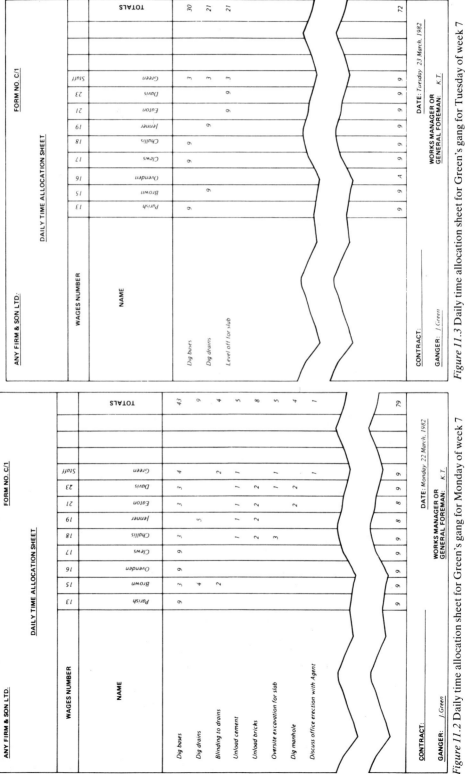

Figure 11.2 Daily time allocation sheet for Green's gang for Monday of week 7

Figure 11.3 Daily time allocation sheet for Green's gang for Tuesday of week 7

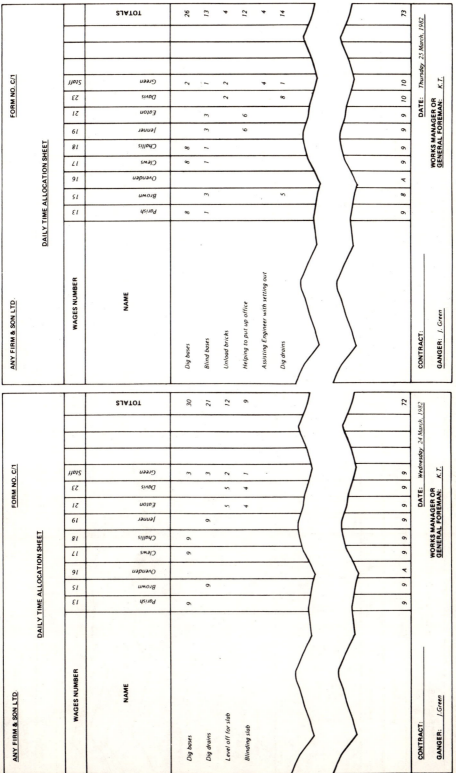

Figure 11.4 Daily time allocation sheet for Green's gang for Wednesday of week 7

Figure 11.5 Daily time allocation sheet for Green's gang for Thursday of week 7

ANY FIRM & SON LTD. FORM NO. C/1

DAILY TIME ALLOCATION SHEET

WAGES NUMBER	13	15	16	17	18	19	21	23	Staff	
NAME	Parish	Brown	Ovenden	Clews	Challis	Jenner	Eaton	Davis	Green	**TOTALS**
Concrete to drains	8					4			3	15
Dig bases	8	4		8						20
Concrete bases 2B & 2C		4			4		4		3	15
Helping to put up office					4					4
Dig drains						4		4	2	10
	8	8	A	8	8	8	8	8	8	64

CONTRACT: _____
GANGER: *J Green*
WORKS MANAGER OR GENERAL FOREMAN: *K.T.*
DATE: *Friday 26 March, 1982*

ANY FIRM & SON LTD. FORM NO. C/1

DAILY TIME ALLOCATION SHEET

WAGES NUMBER	13	15	16	17	18	19	21	23	Staff	
NAME	Parish	Brown	Ovenden	Clews	Challis	Jenner	Eaton	Davis	Green	**TOTALS**
Concrete bases 3B & 3C	3½	3½		3½	3		3	3	3	22½
Unload bricks					½		½	1	1	3
Concrete to bay 1B of ground floor slab	4	4		4	4		4	4	4	28
	7½	7½	A	7½	7½	A	7½	8	8	53½

CONTRACT: _____
GANGER: *J Green*
WORKS MANAGER OR GENERAL FOREMAN: *K.T.*
DATE: *Saturday 27 March, 1982*

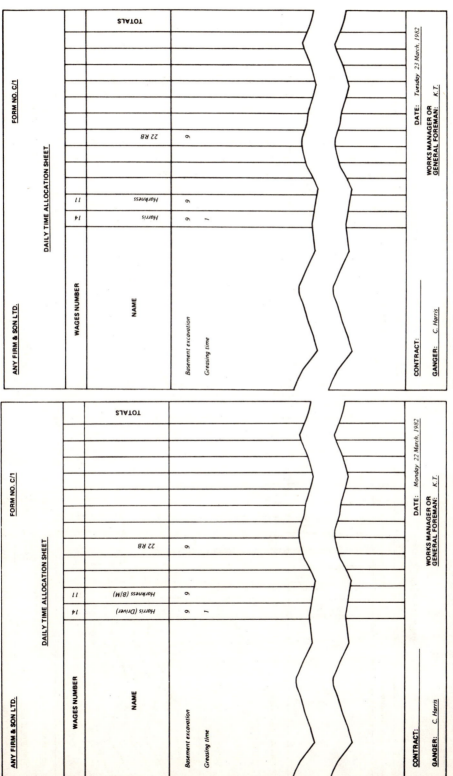

Figure 11.8 Daily time allocation sheet for Harris's gang for Monday of week 7

Figure 11.9 Daily time allocation sheet for Harris's gang for Tuesday of week 7

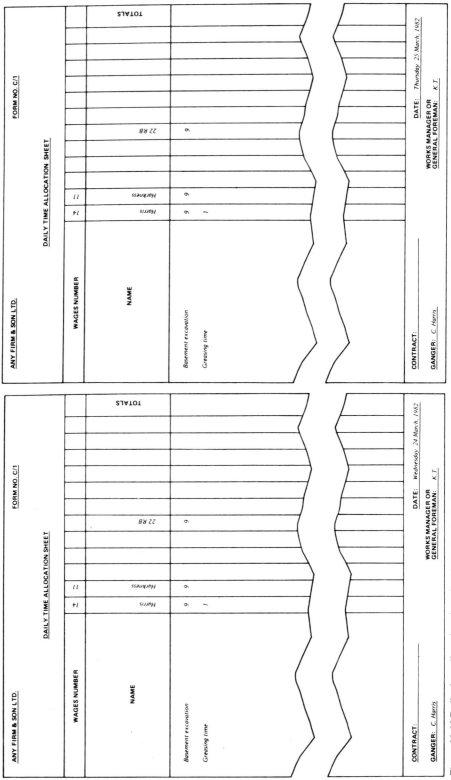

Figure 11.10 Daily time allocation sheet for Harris's gang for Wednesday of week 7

Figure 11.11 Daily time allocation sheet for Harris's gang for Thursday of week 7

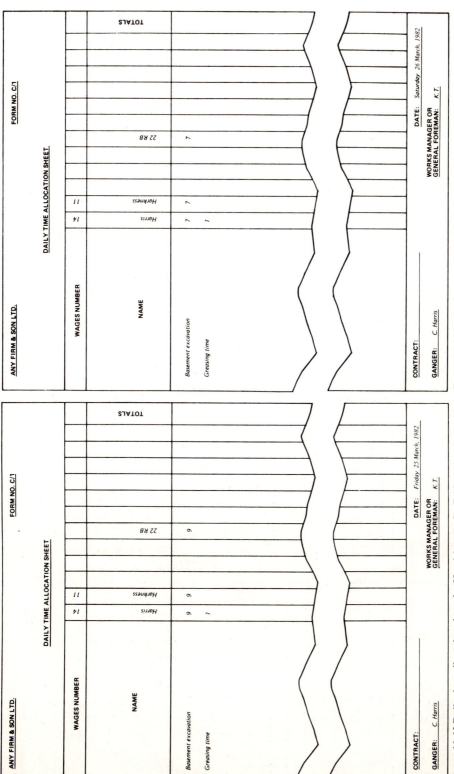

Figure 11.12 Daily time allocation sheet for Harris's gang for Friday of week 7

Figure 11.13 Daily time allocation sheet for Harris's gang for Saturday of week 7

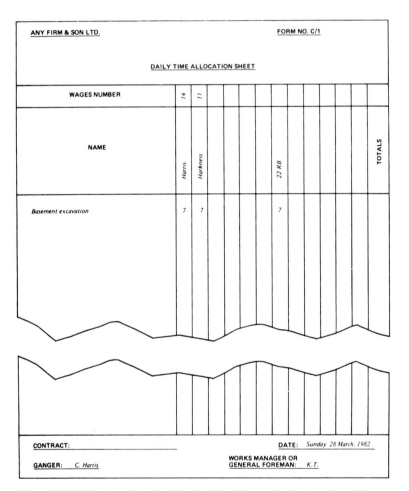

Figure 11.14 Daily time allocation sheet for Harris's gang for Sunday of week 7

ANY FIRM & SON LTD.		FORM NO. W/2
WEEKLY PAYROLL SUMMARY		
BREAKDOWN OF HOURS	DIRECT LABOUR HOURS	WORKING STAFF HOURS
NORMAL WORKING HOURS	359	39
OVERTIME	119½	14
NON-PRODUCTIVE OVERTIME	80½	9
MAINTENANCE TIME	6	—
INCLEMENT WEATHER TIME	—	—
	565	62
ADD WORKING STAFF	62	
TOTAL HOURS	627	

BREAKDOWN OF COSTS	DIRECT LABOUR £p	WORKING STAFF £p
WAGES	918.12	208.00
PLUS RATES	2.48	—
BONUS	172.63	24.12
TOOL MONEY	—	—
FARES	15.20	—
LODGING ALLOWANCE	49.35	—
TRAVEL TIME	10.90	—
SICK PAY	—	—
HOLIDAY STAMPS	90.00	—
GRADUATED NATIONAL INSURANCE CONTRIBUTION	151.27	27.40
REDUNDANCY PAYMENTS	—	—
	1409.95	259.52
ADD WORKING STAFF	259.52	
TOTAL WAGES	£1669.47	

CONTRACT   W/E   28 March, 1982

Figure 11.15 Payroll summary for week 7

W/E 28 March, 1982												FORM NO. C/2
WEEK NO. 7												SHEET NO. 1

QTY.	UNIT	UNIT TARGET	TARGET HOURS	DESCRIPTION OF WORK	M	T	W	T	F	S	S	ACTUAL HOURS
50	m^3	3½	175	Excav. bases	43	30	30	26	20			149
28	m^3	4	112	Excav. drains	13	21	21	14	10			79
10	m^3	4½	45	Excav. oversite	5	21	12					38
7	m^3	5	35	Blinding	4			9	13			26
4	m^3	5	20	Concrete drains						15		15
13	m^3	3½	45½	Concrete bases						15	22½	37½
6	m^3	6	36	Concrete floor slab							28	28
14 000	No.	1½ per thou	21	Unload bricks	8			4		3		15
9	tonne	½	4½	Unload cement	5							5
			18	Erect office	1			12	4			17
			6	Chainman				4				4
			518	TOTALS	79	72	72	73	64	53½		413½

Figure 11.16 Cost collection sheet for Green's gang for week 7

W/E 28 March, 1982								FORM NO. C/2A		
WEEK NO. 7								SHEET NO. 1		
PAY NO.	NAME	M	T	W	T	F	S	S	TOTAL HOURS	BONUS SHARES	BONUS £ p
13	Parish	9	9	9	9	8	7½		51½	52	19.00
15	Brown	9	9	9	8	8	7½		50½	51	18.63
16	Ovenden	9	—	—	—	—	—		9	9	3.29
17	Clews	9	9	9	9	8	7½		51½	52	19.00
18	Challis	9	9	9	9	8	7½		51½	52	19.00
19	Jenner	8	9	9	9	8	—		43	43	15.71
21	Eaton	8	9	9	9	8	7½		50½	51	18.63
23	Davis	9	9	9	10	8	8		53	53	19.37
Staff	Green (Ganger)	9	9	9	10	8	8		53	66	24.12

BONUS PAYMENT

TOTAL HOURS 413½

TOTAL SHARES 429

TOTAL BONUS 156.75

TOTAL TARGET HOURS   518   HOURS
TOTAL ACTUAL HOURS   413½ HOURS
            SAVING   104½ HOURS @ £1.50 = £156.75

            DIVIDE BY   429   NO. OF SHARES
            = £36.54   PER SHARE

Figure 11.17 Bonus share-out for Green's gang for week 7

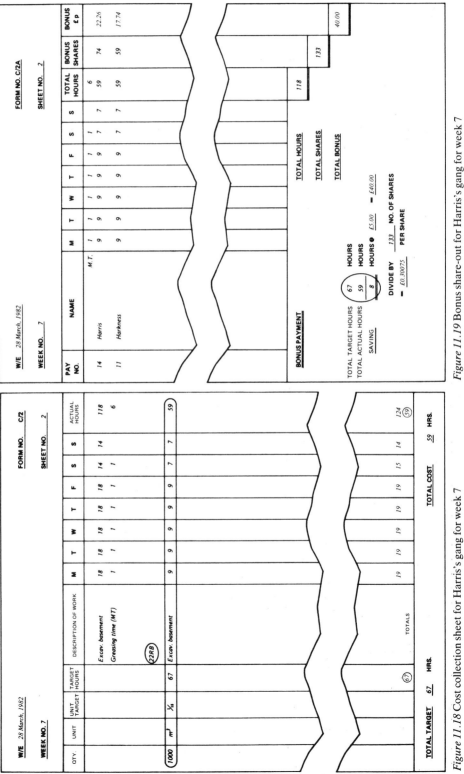

Figure 11.18 Cost collection sheet for Harris's gang for week 7

Figure 11.19 Bonus share-out for Harris's gang for week 7

based on machine hours saved and not on man hours. Throughout the exercise plant and labour costs should be kept separate in order to make possible interim balances with other payroll and plant records. Plant should be shown either in a different colour from labour or separate columns inserted on the sheets. In the illustrations plant is highlighted by enclosing with a circle.

Work carried out by Labour-only sub-contractors should also be kept separate from work carried out by the contractor's own direct labour as this not only facilitates balancing of expenditure but also provides useful comparisons between costs of using sub-contractors and costs of using directly employed labour.

It will be seen that by incorporating the bonus system into the standard cost, duplication of bookkeeping is avoided. Bonus can be calculated as illustrated and the collected data of measurements and expenditure transferred to the feed-back collection sheets as suggested in Chapter 9. If collection of data for feed-back is not required, the information can be transferred directly to a standard cost statement, costing standards being substituted for bonus targets.

In order that a deadline for calculating bonus payments can be met it is essential to abstract allocation sheets daily and pre-measure as much work as possible, so that on measurement day (usually Monday) only the weekend's allocation need be entered, and the odd few measurements taken. By adhering to these principles it is possible to calculate bonuses within only a few hours of the start of a new week, thus avoiding the unsatisfactory practice of paying bonuses one week after wages, i.e. two weeks after the bonus has been earned. Greater incentive effect is obtained by publishing bonus earnings or gang performance levels as soon as they have been calculated.

The next stage is to collect similar work together as shown in Figures 11.20–11.24 and, as they become available, the amounts for Figures 11.25–11.27.

Working down the descriptions on the weekly sheet for J Green's gang, Figure 11.16, the first item to be abstracted is for 50 m^3 of excavate bases taking a total of 149 man hours. This data is entered on the feed-back record sheet, Figure 11.20 for week 7 under the heading of excavate bases. This sheet can also be used as a reminder of relevant facts concerning the item. Bill of Quantity references for instance and the values of £2.25/m^3 for labour and £0.90/m^3 for plant including fuels and consumables. The total quantity measured in the contract can be noted and the rates at which labour and plant are to be costed. Cumulative totals can be brought down each week, preferably in a different colour from the week's figures, there being no entry on excavate bases until week 6 when this item of work commenced. In the illustrations cumulative figures have been underlined to differentiate them from the weekly figures.

Where work is sub-let and is paid for in a manner different from the standard method of measurement appropriate to the contract, then it may be necessary to record two measurements as shown in Figure 11.21 for the item of *cart away* where the value is based on the combined quantities of excavation carried out in any week but the cost is based on the bulk measure agreed with the sub-contractor.

102 *Standard cost example*

The maintenance or greasing time paid to machine drivers for maintaining their machine has, in the exercise, been spread as a cost over all machine excavation items. This cost could, however, be treated as a payroll on-cost and dealt with in the same way as travel time allowance, fares, etc. On the other hand bonus payments have been treated as a payroll on-cost but could have been dealt with as part of the productive cost of carrying out the work and the £1.80 rate per hour used for labour could have been enhanced to cover this. Any item of payroll expenditure can be included with the productive work in this way but it is important especially when comparing data from different sites to be consistent in the definition of what has been taken within the productive cost.

CONTRACT							FORM NO. C/3		
FEED-BACK RECORD SHEET									
Excavator									
REF.	13/A			13/B			25/P — 26/F		
VALUE	L. £1.80	(P. £2.25)		L. £2.25	(P. £0.90)		L. £2.25	(P. £2.25)	
DESC.	Excavate o/s.			Excavate bases			Excavate drains and manholes		
NOTES	B.O.Q. 830	£6.00	£10.00	£1.80	B.O.Q. 650	£10.00	£1.80	B.O.Q. 250	£10.00	£1.80
WEEK NO.	m³	Hours Drott	Hours 22 RB	Hours Lab.	m³	Hours 22 RB	Hours Lab.	m³	Hours 22 RB	Hours Lab.
5	20	10		6 20						
6	100	40	4	10 84	70	10	202			
	120	50	4	120	70	10	202			
7	10	–	–	38	50	–	149	28	–	79
	130	50	4	158	120	10	351	28	–	79

Figure 11.20 Feed-back record sheet up to week 7. General excavation

In deciding this definition, i.e. whether to cost nett or gross as outlined in Chapter 6, or to cost somewhere between nett and gross, consideration must be given not only to expediency but also to how the payroll on-costs will be monitored if they are included against productive work. It may be desirable, for instance, to cost all work at gross payroll rates but to carry out a separate exercise to monitor bonus payments or non-productive overtime. Similar items of payroll on-costs can be grouped together such as public holidays and annual holiday stamps or as suggested in Figure 11.26 all costs associated with the importation of labour on the basis that any expenditure on subsistence would create a saving on travel time and fares and vice versa.

CONTRACT					FORM NO. C/3			
			FEED-BACK RECORD SHEET					
Excavator					Sub-Let			
REF.	13/C				13/D			
VALUE	L. £0.36	P. £0.45			Gross L & P. £0.90			
DESC.	Excavate basement				Cart away			
NOTES	B.O.Q. 3 015	£10.00	£1.80		B.O.Q. 4495	£0.70		
WEEK NO.	m³	Hours 22 RB	Hours Lab.		m³	Sub-Let m³		
5					20	24		
6	100	20	2 40		270	340		
	100	20	42		290	364		
7	1 000	59	6 118		1 088	1 300		
	1 100	79	166		1 378	1 664		

Figure 11.21 Feed-back record sheet up to week 7. Bulk excavation and cart away

CONTRACT									FORM NO. C/3	
				FEED-BACK RECORD SHEET						
Concretor										
REF.	14/A–D & 26/S		14/E		14/H		26/T		14/J	
VALUE	L. £9.00		L. £6.75		L. £10.00		L. £8.10		L. £11.70	
DESC.	Blinding		Bases		150 mm Floor slab		Surround to drains		Columns	
NOTES	B.O.Q. 121	£1.80	B.O.Q. 229	£1.80	B.O.Q. 822	£1.80	B.O.Q. 193	£1.80	B.O.Q. 198	£1.80
WEEK NO.	m³	Hours Lab.	m³	Hours Lab.	m³	Hours Lab.	m³	Hours Lab.	m³	Hours Lab.
6	5	28								
	5	28								
7	7	26	13	37	6	28	4	15		
	12	54	13	37	6	28	4	15		

Figure 11.22 Feed-back record sheet up to week 7. Concrete

CONTRACT							FORM NO. C/3	
FEED-BACK RECORD SHEET								
Labour Prelims.								
REF.	*Bill 1*		*Bill 1*		*Bill 1*		*Bill 1*		
VALUE	£45.00 per week for 20 weeks		£20–£90 per week until £2250		£450 erect £100 dismantle		£180 erect. £80 dismantle		
DESC.	Chainman		Unload materials		Temporary buildings		Mixer set-up		
NOTES	*Value used*	£1.80	*Value used*	£1.80	*Value used*	£1.80	*Value used*	£1.80	
WEEK NO.	£	Hours Lab.	£	Hours Lab.	£	Hours Lab.	£	Hours Lab.	
5	45	12	20	20	180	200	180	90	
6	45	27	20	17	180	180			
	_90_	_39_	_40_	_37_	_360_	_380_	_180_	_90_	
7	45	4	20	20	90	17			
	_135_	_43_	_60_	_57_	_450_	_397_	_180_	_90_	

Figure 11.23 Feed-back record sheet up to week 7. Fixed labour overheads

CONTRACT							FORM NO. C/3	
FEED-BACK RECORD SHEET								
Plant Prelims.								
REF.	*Bill 1*				*Bill 1*				
VALUE	£36.00 per week for 30 weeks				£18.00 per week for 40 weeks				
DESC.	Concrete Mixer				Mortar Mixer				
NOTES	*Value used*	£36.00			*Value used*	£14.00			
WEEK NO.	£	10/7 weeks			£	5/3½ weeks			
5	36	1							
6	36	1							
	_72_	_2_							
7	36	1			18	1			
	_108_	_3_			_18_	_1_			

Figure 11.24 Feed-back record sheet up to week 7. Fixed plant overheads

CONTRACT:						FORM NO. C/3		
FEED-BACK RECORD SHEET								
REF.	Included in Bill of Quantity rates							
VALUE	6.68%		3.21%		3.12%			
DESC.	Non-productive overtime		Inclement weather time		Redundancy and CITB levy			
NOTES	BUDGET USED	£1.80	BUDGET USED	£1.80	BUDGET USED	Redund.	CITB levy	
WEEK NO.		Hours Lab		Hours Lab		£	£	
5	31	25	15	14	14	—		
6	44	23	21	20	21			
	75	48	36	34	35			
7	64	89	31	—	30			
	139	97	67	34	65			

Figure 11.25 Feed-back record sheet up to week 7. Variable overheads

CONTRACT								FORM NO. C/3
FEED-BACK RECORD SHEET									

REF.	Included in Bill of Quantity rates								
VALUE	11·42%		3·55%		14·97%				
DESC.	Holiday Stamps		Public Holidays		Importation of Labour				
NOTES	BUDGET USED		BUDGET USED		BUDGET USED	Travel		Fares	Sub-sistence
WEEK NO.		£		£		£		£	£
5	53	72	16	—	69	6		8	49
6	76	117	24	—	99	14		18	49
	129	189	40		168	20		26	98
7	109	90	38	—	142	11		15	49
	238	279	78		310	31		41	147

Figure 11.26 Feed-back record sheet up to week 7. Variable overheads

CONTRACT:							FORM NO. C/3	
FEED-BACK RECORD SHEET								
REF.	Included in Bill of Quantity rates							
VALUE	1·14 %		33·80%		21·38%		0·86%	1·14%
DESC.	Plus rates and tool money		Bonus		Employer's GNI		SICK Pay	Miscellaneous Payments
NOTES	BUDGET USED		BUDGET USED		BUDGET USED		BUDGET USED	BUDGET USED
WEEK NO.		£		£		£		£		£
5	5	—	156	42	99	98	4	—	5	—
6	8	2	224	83	142	165	6	—	8	—
	13		380	125	241	263	10		13	
7	11	2	321	197	203	179	8	—	11	—
	24	4	701	322	444	442	18		24	

Figure 11.27 Feed-back record sheet up to week 7. Variable overheads

The budgets or values for these payroll on-costs can best be related to the value of work done as discussed in Chapter 10, though some items such as public holidays and inclement weather time, etc., will show gains for most weeks then a sudden loss when expenditure is necessary. This can be safeguarded by only releasing the accumulated budget on weeks when expenditure is incurred. With most overhead items however it is the to-date picture that is more important and for this reason it is useful to record the amount of budget used each week on the feed-back record sheets as indicated in Figures 11.23 to 11.27.

In the exercise payroll on-costs have been taken as simple percentages on the basic rate value of the work done as illustrated in Chapter 10 and calculated in Figure 10.2. Thus in week 7 the £796-worth of measured work and £155-worth of labour prelims shown in Figure 11.30 give a total of £951-worth of work done at basic labour rate level of the payroll. The value or budget of non-productive overtime shown in Figure 11.25 for week 7 is therefore:

£951 × 6.68% = £63.53 say £64

and of inclement weather time:

£951 × 3.21% = £30.53 say £31

and so on, a check being made on the final summary sheet that the total of £968 on-cost value is approximately equal to £951 × 101.27%, i.e. £963, the 101.27% being the total of payroll on-costs shown in Figure 10.2.

Without waiting for the cost totals of these various payroll overheads to emerge from the wage calculations the week's cost summaries can be commenced. Figure 11.28 shows the productive work summarised and followed, when data is available, by Figure 11.29 showing the various site overhead costs. As mentioned above, monitoring of these overheads is often more beneficial on a to-date basis and this sheet can therefore be extended to show both the current week and cumulative data if so desired.

Figure 11.30 summarises all data for the week and to-date on to a single sheet of paper and calculates the overall gain or loss for the contract expressed as a percentage of the value of work done. It is preferable to base such a percentage gain or loss on the value of the work rather than on the cost of the work because the value is the known or fixed side of the comparison whereas the cost is the floating side which is itself being monitored. Presentation of this final summary is most suitable for senior management level of a company where distribution of the detailed build-up of the cost is not required on a regular basis.

The order of working through the various clerical procedures within the system has been so arranged that information is available in the most suitable units with the minimum of effort: first, the calculation of bonus in hours; second, the recording of feed-back information again in hours and tied into the system so that it cannot be left behind; third, the production of the cost statement and summary in cash. By this method the advantages of working both in hours and in cash are utilised.

A check on mathematical error on the expenditure side of the standard cost is provided by a simple reconciliation of wages sheets with total labour

			FORM NO. C/4	
CONTRACT _____		W/E 28 March 82		
SHEET NO. 1		Wk No. 7 of 75		
COST CONTROL STATEMENT				

DESCRIPTION OF WORK	VALUE £	COST £	GAIN £	LOSS £
Excavator				
Oversite excavation	18	68	—	50
Oversite excavation	(22)	—	(22)	—
Excavate bases	112	268	—	156
Excavate bases	(45)	—	(45)	—
Excavate drains & manholes	63	142	—	79
Excavate drains & manholes	(63)	—	(63)	—
Excavate basement	360	223	137	—
Excavate basement	(450)	(590)	—	(140)
TOTAL EXCAVATOR	553	701	137	285
	(580)	(590)	(130)	(140)
Concretor				
Blinding	63	47	16	—
Concrete bases	88	67	21	—
Concrete floor slab	60	50	10	—
Concrete surround to drains	32	27	5	—
TOTAL CONCRETOR	243	191	52	—
SUB-LET				
Excavator				
Cart away	979	910	69	—

Figure 11.28 Cost control statement for week 7. Productive work

			FORM NO.	C/4
CONTRACT _____			W/E 28 March 82	
SHEET NO. 2			Wk No. 7 of 75	
COST CONTROL STATEMENT				
DESCRIPTION OF WORK	VALUE £	COST £	GAIN £	LOSS £
Labour Prelims				
Chainman	45	7	38	—
Unload Materials	20	36	—	16
Temporary buildings	90	31	59	—
TOTAL LABOUR PRELIMS	155	74	97	16
Plant Prelims				
Concrete Mixer	(36)	(36)	—	—
Mortar Mixer	(18)	(14)	(4)	—
TOTAL PLANT PRELIMS	(54)	(50)	(4)	—
Site On-Costs				
Non-prod overtime	64	160	—	96
Inclement weather time	31	—	31	—
Redundancy & training levy	30	—	30	—
Holiday stamps	109	90	19	—
Public holidays	38	—	38	—
Importation of labour	142	75	67	—
Plus rates & tool money	11	2	9	—
Bonus	321	197	124	—
Employer's GNI	203	179	24	—
Sick pay	8	—	8	—
Miscellaneous payments	11	—	11	—
TOTAL SITE ON-COSTS	968	703	361	96

Figure 11.29 Cost control statement for week 7. Overheads

					FORM NO C/5		
CONTRACT_____					W/E 28 March 82		
					Wk No 7 of 75		
	COST CONTROL SUMMARY SHEET						
	THIS WEEK				TO DATE		
	Value	Cost	Gain	Loss	Value	Cost	Gain	Loss
MEASURED WORK								
Excavator	553	701	137	285	963	1357	148	542
Excavator (Plant)	(580)	(590)	(130)	(140)	(958)	(1170)	(153)	(365)
Concretor	243	191	52	—	288	241	52	5
TOTAL MEASURED LABOUR	796	892	189	285	1251	598	200	547
TOTAL MEASURED PLANT	(580)	(590)	(130)	(140)	(958)	(1170)	(153)	(365)
TOTAL MEASURED	1376	1482	319	425	2209	2768	353	912
OVERHEADS								
Fixed labour prelims	155	74	97	16	825	1058	138	371
Plant prelims	(54)	(50)	(4)	—	(126)	(122)	(4)	—
On Costs on £951 value	968	703	361	96	2108	1573	753	218
TOTAL OVERHEADS	1177	827	462	112	3059	2753	895	589
SUB-LET Excavator	979	910	69	—	1240	1165	75	—
TOTAL SUB-LET	979	910	69	—	1240	1165	75	—
GRAND TOTAL	3532	3219	850	537	6508	6686	1323	1501
	Gain/~~Loss~~ This week £ 313 = 8.9 %							
	~~Gain~~/Loss To date £ 178 = 2.7 %							

Figure 11.30 Cost control summary for week 7

costs shown in the standard cost statement. The total should balance to within a few pounds as shown below.

Cost reconciliation

From cost control summary sheet Figure 11.30

	£
Measured work	892
Labour preliminaries	74
On-costs	703
	£1669

From payroll summary Figure 11.15

	£
Workmen's payroll	1410
Working staff	260
	1670
Deduct previous week's bonus if bonus paid one week after wages	N/A
	1670
Add this week's bonus if ditto	N/A
	£1670

Similar checks must be carried out on plant and sub-contract work. It is, however, the labour element of the calculations that is more prone to error because of the numerous calculations involved. Where bonuses are paid one week later than wages, i.e. bonus for week 6 paid with wages for week 7, then an adjustment to the wages summary is needed to apply the correct week's bonus amount in order to achieve a balance. In the above reconciliation such adjustment does not apply as bonus is paid current with wages.

Such reconciliation, first of man hours and now of wages must be carried out at each appropriate stage in the production of the cost statement. The separation of labour, plant and sub-let costs assist in locating mathematical errors. To wait until a grand total is produced before attempting to balance costs is likely to lead to untraceable errors and waste considerable time in searching for possible mistakes somewhere in the calculations.

A further check on mathematical accuracy within the cost statement itself can be made by the simple addition of values and losses compared with costs and gains. For example, grand total for week 7

	£		£
Value	3532	Cost	3219
Loss	537	Gain	850
	£4069		£4069

Standard cost example 113

This particular check can be frequently carried out during the compilation of the standard cost as the rule applies equally to individual operations, trade totals, etc. For example, total plant listed under excavator

	£		£
Value	580	Cost	590
Loss	140	Gain	130
	£720		£720

Exercises

(1) Illustrate your proposals for a standard cost system for a large building firm dealing in hospitals, factories, schools, etc., where the incentive scheme is based on a straight plus rate addition to hourly rates and does not therefore require provision for calculation within the costing system. The statement is to be presented in hours and feed-back of data is to be recorded by a central office on receipt of the standard cost statement.

(2) Design a bonus calculation sheet where work study standards are used as bonus targets. Suggest a means of extending the bonus system to produce a standard cost.

Chapter 12

Standard cost exercise

Exercises

(1) The following prewritten daily task sheets show the work programmed for each day of week 8, and the amount of time spent on each operation. Using the information given on the feed-back sheets previously used for week 7, calculate bonus earnings and prepare a cost statement for week 8.

Measurements for week ending 4 April, 1982 are as follows

Excavate basement and load (22 RB)	1140 m^3
Excavate bases and load	30 m^3
Excavate drains and load	25 m^3
Excavate oversite and load	20 m^3
Blinding	9 m^3
Concrete bases	12 m^3
Concrete slab	11 m^3
Unload bricks	3000 No.

All excavations carted away by labour-only sub-contractor.

CA quantity on lorry 1500 m^3

Bonus targets are as for previous weeks with the chainboy estimated again at 6 h and erecting partitions at 18 h.

(2) Design a control system based on work studied standard times to produce a standard cost statement and an incentive scheme that will pay bonus at the rate of 33⅓% above basic at 100 operator performance (BS 3138:1979 Clause 51034) reducing in steps of 10 performance down to nil bonus at 50 performance. Work through the system using week 8 measurements and the following standard times

Excavate basement and load by 22 RB and dragline (standard time includes for banksman)		0.08 h/m^3	
Excavate bases and load (hand)		2.0 h/m^3	
Excavate drains and load (hand)		2.5 h/m^3	
Excavate oversite and load (hand)		3.0 h/m^3	
Blinding		3.5 h/m^3	
Concrete bases		2.0 h/m^3	
Concrete slab		4.0 h/m^3	
Unload bricks		1.0 h/thousand	
Chain boy	estimated	4	h
Erect partitions	estimated	12	h

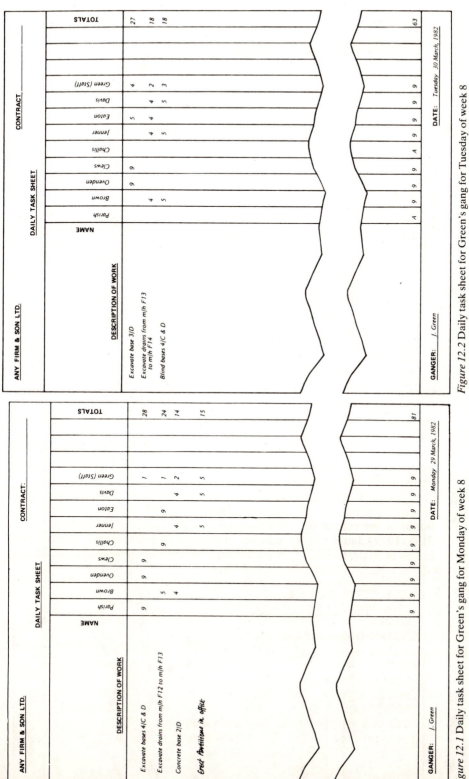

Figure 12.1 Daily task sheet for Green's gang for Monday of week 8

Figure 12.2 Daily task sheet for Green's gang for Tuesday of week 8

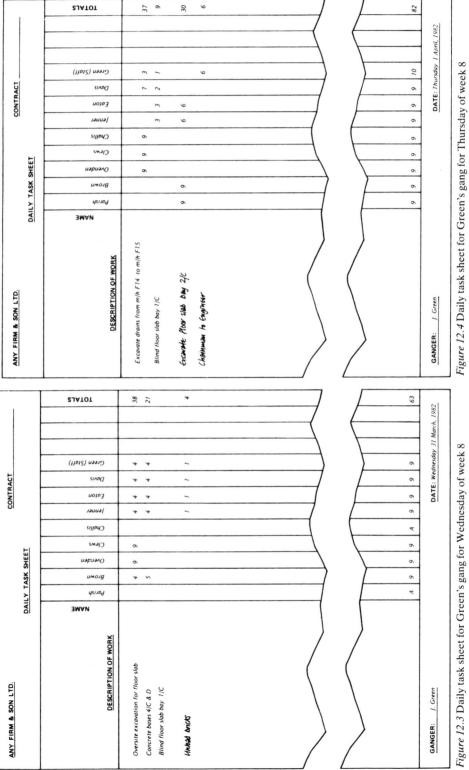

Figure 12.3 Daily task sheet for Green's gang for Wednesday of week 8

Figure 12.4 Daily task sheet for Green's gang for Thursday 1 April of week 8

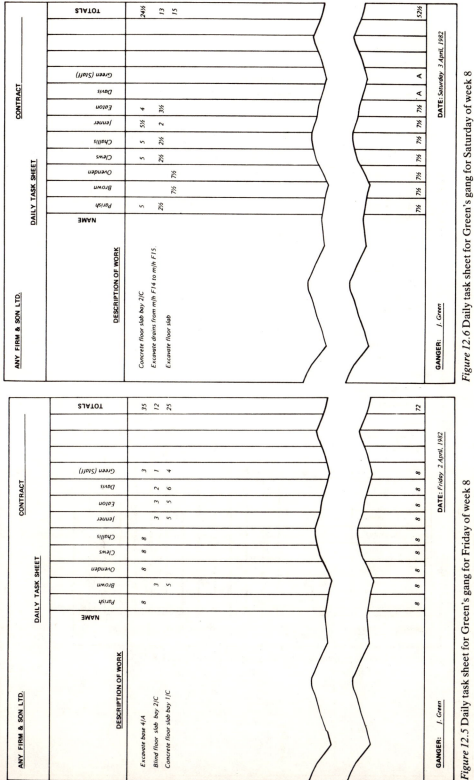

Figure 12.5 Daily task sheet for Green's gang for Friday of week 8

Figure 12.6 Daily task sheet for Green's gang for Saturday of week 8

118 *Standard cost exercise*

ANY FIRM & SON LTD.				CONTRACT				
DAILY TASK SHEET								
DESCRIPTION OF WORK	NAME Harris	Harkness (B/M)		22 RB				TOTALS
Monday								
Basement excavation	9	9		9				18
Greasing time	1							+1
Tuesday								
Basement excavation	9	9		9				18
Greasing time	1							+1
Wednesday								
Basement excavation	9	9		9				18
Greasing time	1							+1
Thursday								
Basement excavation	9	8		9				17
Greasing time	1							+1
Friday								
Basement excavation	8	8		8				16
Greasing time	1							+1
Saturday								
Basement excavation	7½	7½		7½				15
Greasing time	1							+1
Sunday								
Basement excavation	7½	7½		7½				15
	59 +6	58						117 +6
GANGER: C. Harris						DATE: W/E 4 April, 1982		

Figure 12.7 Daily task sheet for Harris's gang for whole of week 8

ANY FIRM & SON LTD.		FORM NO. W/2
TIME & WAGES WEEKLY COST SUMMARY		

BREAKDOWN OF HOURS	DIRECT LABOUR HOURS	WORKING STAFF HOURS
NORMAL WORKING HOURS	358	39
OVERTIME	127½	6
NON-PRODUCTIVE OVERTIME	82½	3
MAINTENANCE TIME	6	—
INCLEMENT WEATHER TIME	—	—
	568	48
ADD WORKING STAFF	48	
TOTAL HOURS	616	

BREAKDOWN OF WAGES	DIRECT LABOUR £p	WORKING STAFF £p
WAGES	923.00	160.00
PLUS RATES	2.80	—
BONUS		
TOOL MONEY	—	—
FARES	15.40	—
LODGING ALLOWANCE	49.35	—
TRAVEL TIME	11.08	—
SICK PAY	—	—
HOLIDAY STAMPS	90.00	—
GRADUATED NATIONAL INSURANCE CONTRIBUTION	150.00	25.00
REDUNDANCY PAYMENTS	—	—
ADD WORKING STAFF TOTAL WAGES		
CONTRACT		W/E 4 April, 1982

Figure 12.8 Payroll summary for week 8

Chapter 13
Costing tips

Small items of labour

In order that a mathematical balance can be achieved and also to ensure that all corners of expenditure are being controlled, it is necessary to include all operations and overheads, no matter how small, in the standard cost system. It is not necessary, however, to keep every operation under a separate heading.

The cost of small items and their respective contributions to the contract's earnings or standard value can be

(1) grouped together under trade headings, the value being either (a) physically measured against known standards, (b) related to the appropriate measured work either as a percentage or a value per unit of measurement, or (c) a previously calculated lump sum allowance each week, or
(2) charged against the appropriate measured work the standard having previously been increased to allow for these small items. (*Note.* Care must be taken not to contaminate feed-back information by charging differing sundry labours, etc., against measured work, thus casting doubt as to what is included.)

Small items that are difficult to allocate, e.g. sundry labours, are best charged against the appropriate measured work, their value having little effect on the standards. Sundry items, however, such as shuttering to nibs and pockets, are best grouped together as suggested in (1) above.

Small items of plant

Light plant can be costed as a grouped item under the heading of plant overheads, expenditure being compared with a budgeted weekly standard or value.

It is not, however, necessary to add together the cost of numerous similar items of plant each week in order to keep a record of expenditure for light plant. Revised totals can be calculated by adding newly hired items or deducting off-hire items as shown in Figure 13.1.

DATE	LATRINES @ 25p/wk £	THEODOLITE @ 50p/wk £	LEVEL @ 40p/wk £	HAMMER @60p/wk £	SAW BENCH @£2.25p/wk £	TOTAL £
7/3/82	0.25	0.50	0.40		2.25	3.40
14/3/82				0.60		4.00
21/3/82			0.40			3.60
28/3/82						3.60
4/4/82	0.25					3.85
11/4/82			0.40	0.60		3.65

Figure 13.1 Method of recording costs of light plant

Credits

Where a profit and loss type of standard costing system is being operated, a contract may well be paid for work not actually carried out, as, for example, with planking and strutting, formwork to bases, etc. If costs have been incurred against this item, such as additional working space, then the item can usually be costed in the normal way. If, however, no expenditure has occurred or it is impossible to identify that expenditure, then the value of the operation can be shown as a credit to the contract and subsequently a straight gain. This should, of course, be indicated as a credit on the cost statement, as illustrated in Figure 7.4 under Formwork to bases.

Stop-ends

The Standard Method of Measurement does not require stop ends to be shown as a separate item in the Bill of Quantities, however, from a costing point of view it is better dealt with as a measurable item, any value for make, fix and strip stop ends being extracted from the concreting or formwork rates where such an allowance is usually made. Failure to do this may result in expenditure being shown against formwork during a particular week with no value because only stop ends had been fixed.

Dayworks

Time spent on daywork operations can be recorded in the same way as normal measured work is recorded, i.e. by task sheets, ganger's allocation sheets, or personal observation. Comparison can then be made by either of the following methods.

(1) Adding or deducting a constant margin to the expenditure. For example, if daywork on labour is priced at + 150%, of which 10% is for planned profit and head office charges, and it is known that variable overheads are costing 125% above nett, then dayworks must be making 15% gain. Therefore, if the cost of daywork during one week was £100, the appropriate value would be £100 + 15% = £115, i.e. £15 gain. This is a rule-of-thumb method and relies on the variable overheads percentage remaining reasonably constant. It is

however, sufficiently accurate to complete the cost statement provided that the daywork does not form an appreciable percentage of the contract.
(2) Where the extent of daywork is significant, it is necessary to abstract the daywork value from the daywork sheets, care being taken to compare like with like, i.e. materials not to be included in value, nett figures to be used if cost system is nett. This requires the daywork system to be constantly up to date, which is, in itself, a worthy bonus. This value is then compared in total with the allocated time spent on dayworks.

Supervision

Supervision may be

(1) direct site supervision – by, for example, gangers and foreman, or
(2) overall supervision – by, for example, agent, engineers, timekeepers.

It may be dealt with in two ways.

(1) By allocating cost of supervision against individual operations. The standards or values must then include an element of supervision to balance this.
(2) Supervision can be shown as a fixed labour overhead and a weekly standard shown in comparison. Staff can be included in this item, although for confidential reasons staff salaries need not be included in the costing system, or can be included as an approximate lump sum so that individuals' salaries cannot be identified.

Direct site supervision lends itself better to allocation than does overall supervision. However, the actual method used for dealing with supervision and the dividing line between direct and overall supervision are not so important provided that consistency is maintained both within the cost statement and in feed-back of output data.

Bonus

In gross costing bonus is included in the average gross rate, but in nett costing bonus remains as a floating item that cannot be set against any known standard.

There are two types of bonus payment.

(1) Attraction money: a policy, spot or standing bonus paid as an incentive to labour to work for a particular company or on a particular site. This kind of bonus is a variable overhead and may well have been allowed for in the calculation of overheads at tender. It should therefore be shown on the cost statement under variable overheads and compared with any value allowed.

(2) A measured bonus system paid as an incentive to better production. This type of bonus attempts to improve outputs and reduce expenditure. The cost of the bonus scheme must therefore be offset by the savings on productive work. This can be done by two methods. (a) If the bonus is not of great significance, it can be shown as a cost with no value to all measured work, either by trades or in total, as illustrated in the standard cost example in Chapter 11 (Figure 11.30). (b) An enhanced average nett rate can be calculated each week to include nett wages plus bonus. Again, this can be done either by trades or in total.

Concrete mixing

Normally the mixing, transporting and placing of concrete can be allocated to the appropriate measurable item of work, but on a large site with a central batching plant this may not be easy or management may wish to study the economics of mixing concrete as a separate operation. This can then be shown as a measurable item under the concrete trade, the standard for mix concrete having been abstracted from the mix, transport and place standards. Mixing mortar for brickwork, etc., can be studied in a similar manner.

Maintenance or greasing time

Maintenance or greasing time can either be dealt with as an overhead or charged against the appropriate measurable item. On a small contract with little plant a separate study of maintenance costs is not necessary. However, on a large contract with, perhaps, a fitting shop or maintenance depot, such items as maintenance time are an important drain on resources and must be studied as a fixed labour overhead compared with any allowance made at tender for maintenance.

General attendance

Theoretically all items of attendance actually out on site can be charged against some item of productive work – for example, picking up cut-offs from shuttering charged to the item of shuttering, cleaning spoil off road charged to the relevant excavation item, etc. Unfortunately this is not so easy in practice. Operations of tidying up site frequently involve a sudden purge or a gang of labourers doing nothing else but keeping a site clean and safe. It is then a difficult task to allocate times against different operations. The simplest solution is to make an allowance for general attendance or clean site under the fixed overheads heading. General items of transport on site, unload materials etc., can be similarly listed and thus prevent feed-back of output data of measurable items being contaminated by attendance items that may vary from site to site.

Piecework

Productive work paid for on a piecework basis either to labour-only sub-contractors or to direct employees has a known value and a known actual cost provided there is no possibility of extras being paid as the work proceeds, e.g. waiting time, allowance for fetching materials, items not included in the price, etc. This work can be costed in the same manner as time-paid employees but by measurement of actual costs instead of labour allocation of costs, as illustrated in Figure 13.2.

DESCRIPTION	MEASURE m^2	UNIT VALUE, £	TOTAL VALUE, £	UNIT COST, £	TOTAL COST, £	GAIN, £	LOSS, £
Formwork to bases	10	6.25	62.50	5.00	50.00	12.50	–
Formwork to columns	60	10.00	600.00	8.75	525.00	75.00	–
Formwork to beams	75	8.75	656.25	8.75	656.25	–	–
Formwork to ribs	4	12.50	50.00	15.00	60.00	–	10.00
Total carpenters			1368.75		1291.25	87.50	10.00

Figure 13.2 Cost statement for work paid for exclusively on piecework

However, where a contract is almost wholly carried out on a piecework basis there seems little point in extending lists of known gains and losses each week simply to show a weekly summary. Office time can be better spent studying costs outside the piecework payments, e.g. dayworks, overheads, variations, etc. If a weekly summary is necessary, an average percentage gain or loss can be calculated for all piecework by trades or sections or in total as soon as rates are known and this percentage can be used to calculate values from the measured weekly payment sheets. For example, if average margin on carpenters for the whole contract is known to be 7% above cost and work carried out this week has cost £258.25, then value this week for carpenters is £258.25 + 7% = £276.33 – a gain of approximately £18, which is sufficiently accurate for general statistical use.

Costing by stages

Certain types of repetitive work are suitable for grouping into stages of construction rather than units of measurement, such as housing or industrialised building. House building can be divided into stages of

(1) excavate footings
(2) concrete footings
(3) brickwork up to DPC
(4) hardcore fill
(5) concrete ground floor slab
(6) first lift brickwork
(7) second lift brickwork
(8) first floor joists
(9) third lift brickwork
(10) fourth lift brickwork
(11) top out brickwork
(12) roof joists

(13) floor boarding
(14) internal partitions
(15) joinery

In addition, there will be any other stages that the contractor intends to carry out himself rather than sub-contract. Drainage, oversite excavation and external works are not easily grouped into such stages and may have to be measured as with a normal standard cost.

Allocation of labour and plant is simplified when made against one of these 15 stages. In practice some of these stages may be sub-let to labour-only contractors and the payment per stage previously agreed, which would simplify allocation even further. The standard or value for

HOUSE NO.	1	2	3	4	5	6	7	8	9	10	11	12
Excavate footings	√	√	√	√	√	√						
Concrete footings	√	√	√	√	√							
Brickwork to D.P.C.	√	√	√									
Hardcore fill	√											

Figure 13.3 Tick sheet for recording completed work on housing contracts

CONTRACT _____

FEED-BACK RECORD SHEET

STANDARD VALUE	£125		£150		£200		£75		
ITEM	Excavate Footings		Concrete Footings		Brickwork to D.P.C.		Hardcore Fill		
WEEK NO.	No.	Hours	No.	Hours	No.	Hours	No.	Hours	
3	2	125							
4	3	184	3	220	1	93			
5	3	202	2	149	2	175	1	36	

Figure 13.4 Feed-back record sheet for housing contract

each stage can be calculated by previous measurement extended at normal unit standards. Weekly measurement can then be in units rather than square metres and can be recorded in a simple tick book, as illustrated in Figure 13.3.

For progress records the ticks can be recorded as a date or in a colour code. Quantities can be abstracted direct from such a tick book to a feed-back collection sheet, as shown in Figure 13.4.

Standing time

Labour

Standing time for labour is best indicated as a cost to a particular section of measured work, e.g. trade or building with no value or standard set against it, thus causing the standing time to 'stick out like a sore thumb' in the loss column of the standard cost statement. It goes without saying that no attempt should be made to hide standing time under measurable work, as this defeats the object of the costing system.

Plant

A certain amount of plant standing is to be expected and can be allowed for, preferably as a plant overhead rather than as an inclusion in output standards. A budget of, say, 20% on all plant on site can then be calculated as the standard value for plant standing and all plant allocated as standing set against this. The difference between standard and actual is of great importance on, for instance, a motorway site. Plant standing time and plant maintenance time can be costed together on smaller sites where they are insignificant.

Apprentices, etc.

Generally the youths or apprentices on site do not warrant special conditions regarding costs and can be treated as fully paid men for costing purposes. However, if this is found to be constantly disturbing the cost balance, such youths can be treated as being a proportion of a man related roughly to their wages scale.

Adjustments

Although small mathematical errors or use of incorrect rates or measurement can be corrected on the next cost statement, it is preferable to show large adjustments as a separate statement; otherwise an error of, say, £100 one week, if corrected the following week, would cause an unrealistic increase or decrease of that £100 in the next cost statement. This may lead site management into research that is unnecessary or may induce a sense of improved profitability that is, in fact, only an adjustment. If feed-back is not being recorded and the cost is being used solely for site control, then adjustments are usually meaningless once the costs have been studied and appropriate action has been taken.

To-date figures

It is helpful to be able to watch the trend of some items, e.g. overheads, and for this reason the cost statement can be arranged to show the 'to-date'

To-date figures 127

CONTRACT							
		FEED-BACK RECORD SHEET					
STANDARD VALUE	£2.25 = 1.25 Hours/m²				OUTPUT —— To Date - - - - WEEK		
ITEM	Fixing soffit shutters including fixing props and span-forms				Standard 1.25 h/m² h/m² 3 2 1		
WEEK NO.	m²		LAB. HOURS £1.80				
12	40		72				
	_40_		_72_				
13	123		150				
	_163_		_222_				
14	211		236				
	_374_		_458_				
15	194		228				
	_568_		_686_				
16	208		231				
	_776_		_917_				

Figure 13.5 Graphically presented output record

or 'last-month' figures alongside this week's statement. Alternatively, key items can be plotted graphically either on a separate form or as an extension to the feed-back file, as illustrated in Figure 13.5.

Graphical controls

A wide variety of graphs, histograms and charts can be used to show pictorially the financial position of the various elements of a contract.

Figure 13.6 shows a histogram of values and costs for a contract set against the programmed value of work. This not only shows gains or losses but also provides an indication of general progress. Such a histogram can be used to show all work or individual trade totals. Values and gains can be shown in one colour, costs and losses in another colour. Labour, plant and consumables can be shown together or as a separate chart. The decision on which items to show separately rests largely with the type of contract and likely trouble spots on the site. Another important factor is the reaction to quantitative detail by site management, where a standard cost statement may make little impression, a graphical presentation may have the desired effect.

Figure 13.7 shows a break-even chart used to calculate the average weekly concrete output required to break even on the cost of mixing concrete.

128

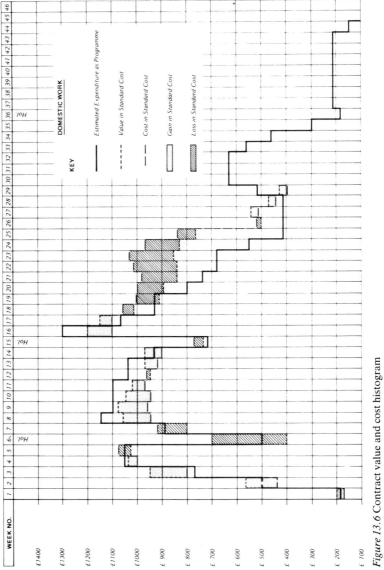

Figure 13.6 Contract value and cost histogram

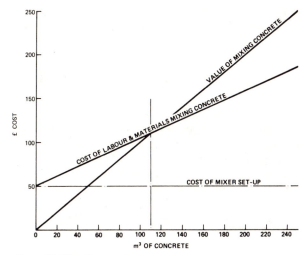

Figure 13.7 Break-even chart

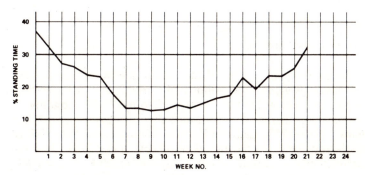

Figure 13.8 Maximum/minimum graph

Figure 13.8 shows a maximum/minimum graph on which standing time for concrete transporters has been plotted. If the graph shows above the maximum line, management must check whether a transporter can be sent off site; likewise, below the minimum line management must check whether an additional transporter is needed.

Other suggestions for graphical records are

(1) value or number of materials on site. An excess of materials on site leads to increased wastage, whereas too small a stockpile leads to hold-ups in programme
(2) number of men on books each week compared with number programmed
(3) percentage of absenteeism
(4) ratio of staff salaries to measured work
(5) percentage of pay-offs each week
(6) cubic metres of concrete placed each day or week; cubic metres of excavation or number of bricks, etc.
(7) wastage percentages of materials

(8) comparison of trade gains or losses so that common changes in profitability can be related to weather, holidays, illness, changed conditions, etc.

Effect of standards on variable overheads

As the value of variable overheads varies directly with the value of measured work, it follows that an exceptional gain or an exceptional loss on measured work will affect the margin of gain or loss on the variable overheads, thus accentuating the results of the measured work on the overheads. Figure 13.9 shows the final summary of measured work and fixed overheads, and of variable overheads where the total value of

DESCRIPTION	VALUE, £	COST, £	GAIN, £	LOSS, £
Measured Work and Fixed Overheads	2000	1500	600	100
Variable Overheads	1200	1150	50	—

Figure 13.9 Effect on variable overheads of measured work

variable overheads adds up to 60% of the total value of measured work plus fixed overheads. A gain of £50 is shown on variable overheads with no losses. However, the reason for this gain is not so much the economy on the variable overheads as the 25% gain on value of measured work and fixed overheads. If this 25% gain were to be removed and the value of measured work and fixed overheads taken at cost, i.e. £1500, then the value of variable overheads would be much less, i.e. £1500 × 60% = £900. Compared with this imaginary figure, the variable overheads would now show a loss of £1150 less £900 = £250, which obviously requires further research and would have remained hidden if such an imaginary calculation had not been made. It is therefore advisable always to check how profitable on-costs would have been if the measured work were running at par.

Spot costs

The detail shown in the standard costing system may not be sufficient to indicate the true reason for a loss on a particular operation or particular section of work. A special cost study or spot cost is therefore required of this item, splitting operations down into more detail: for example, tradesmen and labourers kept separate; temporary works such as barrow runs, templates, etc., kept separate; various elements of the operation kept separate – fetch materials, cut to size, position, fix, clear up. The spot cost needs to give full details of these cost elements on the spot cost statement and must not be summarised on to a separate form for presentation to management, as this would defeat its purpose. Figure 13.10 illustrates a

CONTRACT _____   SHEET NO. _____

DESCRIPTION *Construct Pump Plinth*

DATE	DESCRIPTION OF OPERATION	TYPE OF LABOUR	HOURS	RATE £	AMOUNT £	TYPE OF PLANT	HOURS	RATE £	AMOUNT £
5 July 82	Clean off area of plinth	lab.	2 × 4	3.50	28.00	2T Comp.	4	8.00	32.00
	Bush hammer floor slab	"	2 × 4	3.50	28.00	2T Comp.	4	8.00	32.00
	Make shutters	carp.	1 × 5	4.00	20.00				
6 July 82	Fix shutters to plinth	"	2 × 2	4.00	16.00	Mobile crane	2	20.00	20.00
	Prop, line and level shutters	"	2 × 3	4.00	24.00				
	Make boxes for pockets	"	1 × 4	4.00	16.00				
	Fix boxes to shutter sides	"	2 × 3	4.00	24.00				
7 July 82	Hang boxes in plinth top	"	2 × 4	4.00	32.00				
	Line and level boxes in plinth top	"	2 × 4	4.00	32.00				
8 July 82	Place ready-mix concrete in plinth	lab.	4 × 3	3.50	42.00	Poker vib	3	2.00	6.00
	Joiner standby during concreting	carp.	1 × 2	4.00	8.00	Mobile crane	3	20.00	60.00
	Trowelling	lab.	1 × 1	3.50	3.50				
9 July 82	Strip shutters	carp.	2 × 1	4.00	8.00	Mobile crane	1	20.00	20.00
	Strip boxes	"	2 × 3	4.00	24.00				
	Rub up concrete	lab.	1 × 1	3.50	3.50				
				Labour	309.00				
				Plant	170.00				170.00
				Total Cost	£479.00				

Figure 13.10 Spot cost

spot cost for an operation of concreting a plinth that is known to be additional to contract and for which no similar rate exists in the Bill of Quantities. The hours have been cashed out gross in this instance, i.e. to include all items on the payroll. The spot cost will assist the quantity surveyors to build up a rate for this item.

Allocation of labour and plant would normally be more carefully controlled than under a standard costing system because only a small section of work is being studied. Therefore personal attention can be given to the spot cost allocation.

Spot costs can also be used as a check on the standard costing system to prove the accuracy of the allocation system, measurements, bookkeeping, use of correct standards, etc.

Where no standard costing system is in use, a spot cost on selected items can be used to ascertain the profitability of key operations and thus provide some degree of control.

Marginal costs

Marginal costing is the consideration of only the variable costs of producing an item of work. For example, if a site has a batching plant for producing concrete, the marginal or additional cost of mixing each extra cubic metre of concrete would consist only of the immediate costs of producing that concrete, ignoring all costs for erection and dismantling of the batching plant and hire of equipment, all of which would be a cost to the site whether or not the extra cubic metres of concrete were produced.

Exercise

(1) Daywork has been valued in the costing system of a complex town centre contract at 10% above cost but a thorough spot check for one week comparing allocated expenditure of labour, plant and materials for daywork operations with the amount of daywork submitted to the client gave the following results

Expenditure	£120
Daywork submitted	£90

Therefore loss on daywork £30 or 33⅓% of value. Cost statement was showing £120 as cost and £132 as value, i.e. a gain of £12. Apparently expenditure is being missed at some point in the daywork system. State what the danger points are and what action you would take to prevent this happening in future.

Chapter 14

Costing of jobbing works

For the small or jobbing builder many of the procedures suggested in the previous chapters will appear to be taking a sledge hammer to crack a nut. The smaller and shorter the contract the more a simplified approach to cost control is necessary. It is nevertheless just as important to know how each contract fares compared with the quotation given and many of the principles discussed in the previous chapters will apply. The company's books must certainly be capable of allocating expenditure against individual contracts as described in Chapter 2.

As with the large contractor the biggest problem lies in the allocation of the cost of resources to individual jobs. This problem is, however, even more acute with the smaller contractor who will usually have only one central payroll or even resort exclusively to labour-only sub-contractors. His men will, for instance, be on one contract on Monday to move scaffolding to a different contract on Tuesday, returning with the company's excavator to the first contract on Thursday afternoon. It is very easy to lose track of what resources should be charged to which contract.

Figure 14.1 suggests a simple *job cost sheet* for the foreman or other responsible person to record all resources that are used on a job during a particular week. The rates can be entered and cashed out in the office later if not done on site. Labour rates should include all payroll on-costs such as holiday stamps and Graduated National Insurance and plant rates should include all fuels, oils and consumables if these are not taken as overhead or material items of cost.

Overheads are added in whatever form the contractor has chosen to apportion them to different jobs. In the example they have been assessed as a percentage of total site expenditure. If the builder had overheads for his yard and office of say £15 000 per year and he spent £100 000 a year on wages, plant and materials, then he needs to add 15% to these costs to cover his overheads. If the job is a small one for which no firm quotation has been given, then it only remains to add profit, if this has not already been included in the various rates, and the weekly job cost sheet can then be used as the basis of an invoice. If, however, a quotation for the job has been given and accepted then the weekly job cost sheet acts as the basis of a comparison against that quotation.

133

134 *Costing of jobbing works*

Figure 14.2 illustrates such a comparison summarising each week's costs from the weekly job cost sheet and comparing each element of the quotation, i.e. labour, plant, materials, overheads and profit with those costs. The accepted quotation is shown as £3550 of which £1400 is allowed for labour, £540 for plant and £900 for materials, a total anticipated site

JOB NO 82/3/151			WEEKLY JOB COST SHEET					W/E 4 April 82			
LOCATION 18 The Avenue, Low Town											
JOB Loft conversion											

NAME OF MAN	TRADE	MON	TUES	WED	THURS	FRI	SAT	SUN	TOTAL	RATE	£	P
R. Smith	Joiner			9	9	7	—		25	4·00	100	00
J. Carter	"			4	9	7	7		27	4·00	108	00
R. Baker	Labourer			9	9	7	7		32	3·50	112	00
TOTAL LABOUR				22	27	21	14		84		320	00

TYPE OF PLANT	MON	TUES	WED	THURS	FRI	SAT	SUN	TOTAL	RATE	£	P
8 ton tipper lorry				8	4			12	12·00	144	00
10 cwt van (per day)			1	1	1	1		4	15·00	66	00
Power saw (per week)								1	1·00	1	00
TOTAL PLANT										205	00

DETAILS OF MATERIALS	QUANTITY	UNIT	RATE	£	P
Timber as Smith Bros invoice No 6235/931				294	20
2 rolls roofing felt ex yard	2	No	8·50	17	00
3 rolls fibreglass insulation DIY Ltd receipt No 0321	3	No	5·00	15	00
1 pack 50 mm nails ex yard	1	No	1·00	1	00
				327	20

TOTAL LABOUR PLANT AND MATERIAL COSTS	852	20
OVERHEADS 15 % ON ABOVE	127	83
PROFIT not applicable % ON ABOVE	—	—
TOTAL ALL COSTS	£ 980	03

Figure 14.1 Weekly job cost sheet

expenditure of £2840. 15% on this expenditure has been allowed for overheads, i.e. £426 and 10% for profit, i.e. £284.

During week ending 4 April, 1982 the weekly job cost sheet shows that £320 has been spent on labour. The site foreman or his superior assesses the percentage of work completed in the job at 25% which gives a budget so far at 25% of £1400 – £350. Thus the expenditure so far on labour

JOB NO 82/3/151	JOB COST SUMMARY	DATE OF QUOTATION 8 March 82
LOCATION 18 The Avenue, Low Town		
JOB Loft conversion		
ACCEPTED QUOTATION £ 3550		DATE OF ACCEPTANCE 23 March 82

TOTAL LABOUR BUDGET £ 1400

UP TO WEEK ENDING	% COMPLETE	BUDGET	COST	GAIN	LOSS
4-4-82	25 %	350	320	30	—
11-4-82	75 %	1050	1040	10	—
18-4-82	100 %	1400	1375	25	—

TOTAL PLANT BUDGET £ 540

UP TO WEEK ENDING	% COMPLETE	BUDGET	COST	GAIN	LOSS
4-4-82	33⅓ %	180	205	—	25
11-4-82	66⅔ %	360	395	—	35
18-4-82	100 %	540	585	—	45

TOTAL MATERIALS BUDGET £ 900

UP TO WEEK ENDING	% COMPLETE	BUDGET	COST	GAIN	LOSS
4-4-82	40 %	360	327	33	—
11-4-82	90 %	810	775	35	—
18-4-82	100 %	900	870	30	—

TOTAL OVERHEADS BUDGET £ 426

UP TO WEEK ENDING	% COMPLETE	BUDGET	COST	GAIN	LOSS
4-4-82	890/2840	133	128	5	—
11-4-82	2220/2840	333	331	2	—
18-4-82	2840/2840	426	424	2	—

Actual profit = Gains − Losses + profit anticipated

TOTAL ANTICIPATED PROFIT £ 284

UP TO WEEK ENDING	% COMPLETE	PROFIT ANTICIPATED	ACTUAL PROFIT	TOTAL GAINS	TOTAL LOSSES
4-4-82	890/2840	89	132	68	25
11-4-82	2220/2840	222	234	47	35
18-4-82	2840/2840	284	296	57	45

Figure 14.2 Job cost summary

appears to be within the budget at least to the degree of accuracy that can be expected from such an assessment. As the job progresses so the percentage assessed will become more accurate until 100% completion is reached when the true figures will be known.

The various elements of the work are best kept separate as shown in the example both for assisting with assessing more accurate percentages

WORKMAN'S WEEKLY RECORD CARD

NAME: J. Carter   PAY NO: 14   TRANSPORT: 10 cwt van   W/E: 4 April 82

JOB REFERENCE	HOURS ON JOB INCLUDING TRAVEL TIME								RATE	TOTAL	
	MON	TUES	WED	THURS	FRI	SAT	SUN	TOTAL		£	p
82/3/143 New windows 24 Poole St. High town	8	8						16	4·00	64	00
82/4/165 Chimney pots Mrs Jones, Oaks crescent			4					4	4·00	16	00
82/3/151 Loft conversion 18 The Avenue, Low Town			4	9	7	7		27	4·00	108	00
TOTAL LABOUR	8	8	8	9	7	7		47		188	00

DETAILS OF PLANT AND MATERIALS MOVED BETWEEN JOBS OR THE YARD

MAKE AND MODEL OF PLANT	MOVED FROM:	MOVED TO:	DATE	TIME
Power saw ref S.23	Poole St.	The Avenue	31 March 82	not applicable

TYPE OF MATERIAL	QUANTITY	MOVED FROM:	MOVED TO:	DATE	RATE	TOTAL	
						£	p
Chimney pots	2 No	yard	Oaks crescent	31 March 82	5·00	10	00
Cement	½ bag	The Avenue	Oaks crescent	31 March 82	3·00	1	50
Sand	1 sack	yard	Oaks crescent	31 March 82	say	0	50

Figure 14.3 Workman's weekly record card

complete and also to highlight which elements of the quotation are making or losing money, e.g. that savings on labour are not subsidising wastage on material, etc. A broader comparison can be made, however, by ignoring this breakdown and simply comparing the weeks' total cost with the proportion of the quotation completed so far.

As the weeks progress so the various costs are added to the job cost summary to give an updated picture of the state of the job. Each week the gains and losses are totalled, week ending 4 April, 1982 showing total gains of £68 and total losses of £25. These figures, however, do not include the profit anticipated in the quotation as this element of the budget has so far been ignored. Of the £2840 anticipated site expenditure, only £890 has so far been spent, therefore it is logical to say that 890/2840 of the £284 anticipated profit, £89, has so far been realised. This can therefore be added to the gains and losses actually made on the contract in order to calculate the current actual profit level,

£68 Gain − £25 Loss + £89 Anticipated profit
= £312 Actual profit

Where jobs are so small that no foreman is responsible for a particular site and supervision is as itinerant as the workforce then only the workmen themselves are in a position to know exactly what their movements have been in any particular week. An alternative record may then be necessary revolving not around the job but around the workman himself. Such a record is suggested in Figure 14.3 showing the week's movement for J Carter and details of any plant or materials with which he has been associated so that the office can pick up these movements and transfer any charges where appropriate. Any time spent on a larger contract where a foreman is available must still be recorded by J Carter at least until the end of that week so that the records can be cross-checked with the data entered by the foreman in Figure 14.1. Duplication of such records is always preferable to gaps in the system with all that entails in trying to find out where an item of plant or a couple of labourers were for a few days.

To assist in the correct completion of either the weekly job cost sheet or the *workman's weekly record card*, job headings or references can be entered before issuing so that there is less chance of a job being referred to by an incorrect title.

Part 2
Cost improvement

Motto

Cost control is judged not by the statistics collected but by the action arising from those statistics.

It requires a saving of only five minutes per man per day to achieve an improvement in productivity of 1%.

Chapter 15

Introduction to cost improvement

Action

Undoubtedly many producers of cost control data breathe a sigh of relief when they finally balance their cost statements and copy them for distribution to their superiors. This, however, is not where cost control finishes but where it starts. The data are not yet cost effective and never will be unless action is now taken on the data produced.

Work study

In order to examine jobs of work more closely, some in-depth study of the work may be necessary using work study techniques.

BS 3138:1979 defines work study as 'The systematic examination of activities in order to improve the effective use of human and other material resources'. Such a definition embraces the whole field of cost control; however, the term work study is more commonly used to refer to the use of the various techniques of work measurement and method study as illustrated in Figure 15.1.

The techniques of *work measurement* such as time study, activity sampling, etc., deal with the timing of a particular task, whereas the techniques of *method study* such as multiple activity charts, flow diagrams, etc., deal with improving the way that task is carried out.

The technique of *random activity sampling* is easily learned and can quickly help site management to home-in on problem areas. However, the techniques of *time study* and *multiple activity charts* are probably the most useful in the construction industry enabling the work to be accurately timed and then balanced to produce either improvements in output and/or reductions in cost.

A particularly useful machine for producing multiple activity charts is the TYMLOG recorder which avoids time consuming stop watch work with all the resultant calculations that have to follow a conventional time study.

A good example of how detailed such work study data can become is produced by the Federation of Swedish Building Employers which lists

142 *Introduction to cost improvement*

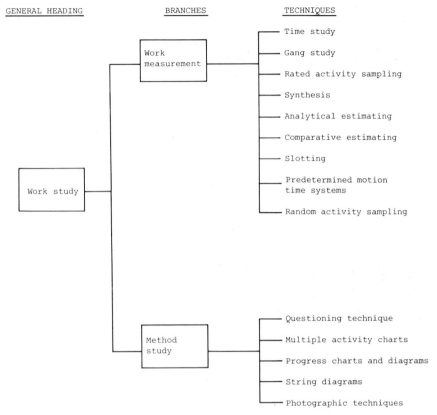

Figure 15.1 Branches and techniques of work study

each job of work on three schedules – blue sheet containing details of method, brown sheet containing time standards and yellow sheet giving a detailed job description. Preferred methods are marked accordingly. The Swedish system of payment by locally agreed piecework rates was rationalised in 1934 to form a national list and is now superseded by rates based on this work study data. Similar systems of work study data banks exist in Germany and in Holland.

Evaluating options

Many decisions on a construction site can be assisted by costing out the options, not only at the start of a contract but also during the course of the works to see that the initial decisions still hold good. Cost comparisons are illustrated for

(1) bulk excavation by scraper or by excavator and lorries
(2) site-mixed or ready-mixed concrete
(3) cut and bend reinforcement on site or purchase ready
(4) hours to be worked on site
(5) recovery of national increases

Incentives

No matter how much effort is put into improving working methods with the resultant gain in productivity, the performance of the individual workman may not change. In order to further improve that productivity some form of motivation of the workman himself is necessary. A well-run incentive scheme can provide that motivation.

Job cards

One of the principles of a good incentive scheme is that the workmen should have visible targets. One way of achieving this is to produce a card for each job of work showing details of the work to be done and the bonus target for that work.

Weekly planning

It is of little use setting targets for an incentive scheme if work is not properly planned in sufficient detail to assist the workforce to flow rapidly from one job of work to the next. Weekly planning meetings are necessary to think through the following week's work not so much to decide *what* is to be done – this may be obvious anyway – but to decide *how* that work is to be done.

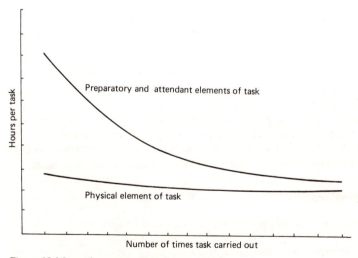

Figure 15.2 Learning curve effect (not to scale)

Figure 15.2 illustrates that a learning curve is caused, not so much by the physical work itself but by the preparatory and attendant work such as discussion about the job, looking at drawings and work sheets, taking measurements, collecting tools and materials together, waiting for other members of the gang to complete their part of the work, weighing up alternative methods, giving and receiving instructions, etc. The more that this preparatory work can be reduced, the higher the chances of improved productivity.

Pre-costing

One of the major problems with post costing, i.e. costing after the work has been carried out, is that many of the jobs will never be repeated and therefore little can be done to recover a loss even if the causes of those losses have been exposed. By assessing values and costs of future weeks' work some attempt can be made at evaluating losses before they occur whilst time still remains to take corrective action.

Computer costing

Study of costing procedure will show that production of cost data is little more than the manipulation of virtually the same sets of figures into an endless variety of outputs, performances, ratios, costs, etc. This manipulation is clearly a candidate for computer assistance. Before this statement can be developed it is essential to understand a little about modern computers in order to appreciate both their potential and their restrictions.

Like them or not, computers will play an ever increasing part in business and will make possible the monitoring of contracts and the evaluation of alternative methods of working that hitherto would not have been cost effective to set up. The danger of distant main-frame computers churning out meaningless data will always be with us but that danger will diminish with the increasing use of interactive processing either as microcomputers or as local terminals where contract management has more direct control over what the computer does with the data and can interact with the computer to produce intelligent results that neither man nor machine could produce alone in the time.

The introduction of computers on sites will not, however, happen as an overnight miracle relieving the drudgery of endless paperwork at a stroke. The technology is already available but the application of that technology will be governed by the investment in both time and money that construction managers are prepared to devote to the development of programs for the construction industry.

Chapter 16
Action

Many facts shown up by the site costing system will, in themselves, suggest a possible remedy – loss on staff, over-cementing of concrete, high percentage of plant standing, etc. However, the majority of operations or overheads will still require further research. Whichever way a cost statement has been calculated, it is no more than a statement of *where* time and money have been expended and *where* losses have been made. Action must now be taken to find out *why* such losses have occurred and *what can be done* to put them right.

Reasons for losses can be summarised under such headings as, Site inefficiency, Interference by client and Estimator's outputs.

Site inefficiency

This may of course be due to incompetence, but is more likely to be the result of lack of communication, lack of planning, lack of materials, lack of suitable labour or lack of incentive. These possibilities must be studied and the root of the inefficiency discovered. Comparisons between gangs of men or similar sites may indicate a reason. Perhaps a bonus scheme is required; perhaps stores are not adequately controlled or lines of command not well defined. Short-term planning can help to sharpen up on management decisions. Job cards can assist in the flow of work. Pre-costing can help to set visible targets. Spot costs can make a more detailed study of a particular operation. Work study can isolate delays and interruptions and calculate method improvements.

It is helpful to further break down losses under this heading into, for example, the following categories

(1) weather
(2) lack of continuity of work
(3) lack of materials
(4) excess labour on section
(5) making good
(6) quantities in excess of drawings

Interference by client

Many alterations, additions, deductions or changes may occur during the course of the contract, over which the contractor has no control. All such alterations should be charged to the client, as they were not envisaged in the estimate.

Losses must be checked to see whether they have occurred as a result of such an alteration and the machinery set in motion to recover excess costs either by: (a) daywork, (b) increased rate, (c) increased measure or (d) claim.

Estimator's outputs

Where the standard cost uses the estimate for calculation of standards, an error or misjudgement by the estimator may cause an incorrect standard or value to be used in which case a loss must be accepted. As feed-back of outputs to a standards library or to the estimator increases, so this kind of error should disappear.

Action sheet

In order to obtain further information from gangers, foremen, etc., as to the reasons for losses, an action sheet is advisable similar to the example illustrated in Figure 16.1. The more important losses are thereby put to those closest to the operation and their comments either filled in by them or made verbally and entered in the Action taken column by a production surveyor, work study engineer, etc. This then gives site management the lead necessary to categorise the losses and take measures to recover costs or prevent further losses of the kind.

Management check-list

A good manager will not wait for losses to occur on his site but will make time available to check up periodically on all functions for which he is responsible. It is useful in this respect to prepare a management check-list at the commencement of a contract noting briefly all the checks that must be made from time to time from the stock of cement to the need for keeping the concrete mixer on site. A time interval then needs to be added to each item bearing in mind that it may not be the manager's direct responsibility to, say, order cement, but it is his responsibility to check that his subordinate has a foolproof system for ordering it.

The format and detail of such a check-list will depend on the personality, experience and time available of the manager himself but to ignore such an *aide-mémoire* will leave the manager open to being drawn from one problem to another without time to take any preventative action before the next crisis arises. Decision making is often considered to commence with

CONTRACT				DATE	
	ACTION SHEET				
DESCRIPTION OF WORK	VALUE £	COST £	LOSS £	ACTION TAKEN	
Concrete base to crane beam	45	150	105	This beam had to be concreted in phases to provide access for other contractors. Q.S. please note	
Shutter base to temperature measuring post	10	60	50	¼ mile from joiner's shop. Would be better recorded as daywork	
Cutting steel with burning gear	—	75	75	Making good. Domestic loss	
Excavate compressed air line trench	160	240	80	Should have been measured through existing road. Record note now prepared	
Backfill land drain with stone	350	525	175	Appears in order but method study being carried out to see if any improvement can be made	
Strip shutter from gate post base	20	60	40	Shutter props part buried by our own temporary access road. Domestic loss	

Figure 16.1 Weekly action sheet

identifying the objectives of the problem but a good manager will start to act even before the problem arises by having an awareness of opportunities that may or may not arise and thus will not be in the situation of having to make a decision under duress.

Getting the idea across

The means of presentation of cost data can have an enormous bearing on the amount of action taken on that site. First, the site or contracts manager must play his part by seeing that the facilities are available to produce the cost control system in the way that it has been designed. So often the will is there to produce copious statistics and regular feed-back of data to a variety of interested parties within a company but the job of producing it is tacked on to the duties of some unfortunate person who cannot possibly be expected to do justice to such a time-consuming task.

Second, it is the duty of the production surveyor/work study engineer to evaluate the work involved in producing his cost reports and to bring to the attention of his superiors any likely holdups in the system so that compromises can be reached between what is desirable from the cost system and what is realistically achievable. Speed is essential in the production of any kind of cost data in order that action can be taken. If, for instance, only three-quarters of a weekly cost report is obtainable then it is better to circulate an interim statement than to wait a couple of days until the final figures can be inserted, balanced and typed.

It is prudent to be critical when being asked for cost information. People are often not precise in what they want to know. It is easier to be vague when requesting cost data and then be more specific when an exercise has been completed and some figures are available to prompt further questions. Such statements as 'Do a cost on those concrete bases' may lead to 'I want to know why the batcher is only working at 50% capacity?' or 'The brickwork costs are high, what kind of output are we getting?' may end up as 'Why have we go three labourers to every fiver bricklayers?'. Much time can be saved in anticipating such subsequent questions arising from the answer to the first and a good surveyor will set up his research so that he can draw-off a number of facts without having to reabstract all his figures.

It is of course prudent to know the background to all figures produced and to assess their relative reliability. It is no use constructing a house on a foundation of sand just as it is no use calculating costs, ratios, outputs and a variety of far-reaching statistics on, for example, a basis of measurements that have been 'paced out' by the chargehand on site.

Of utmost importance is the definition of what is included in a cost, basic rates only; basic and bonus; total payroll; payroll and plant, including or excluding fuels, oils and consumables; including or excluding site overheads, head office overheads, planned profit, etc. Anyone working in management services, i.e. work study engineer, cost or production surveyor, will appreciate the difference between knowing that something is wrong and getting someone to do something about it. The solution may not be so obvious to higher management but unless a good job is made of selling the solution, action will not be taken. On the other hand it is management's duty to pry into any proposals put forward for improving costs and not discard them simply because 'the report runs to 30 pages with no conclusion' or 'the idea sounds like a criticism'.

It has already been stated that speed is imperative in the production of cost data. Putting requirements into perspective will help to generate that speed. Rounding figures off to the nearest pound or even hundred pounds depending on the problem in hand, using averages such as an average cost per man hour, grouping of similar items etc., will all help to produce the facts that few hours earlier and equally, will enable action to be taken sooner.

It must be considered, however, that effecting improvements is not a finite activity. The method will never reach perfection and in the construction industry the job will often be completed before a thorough study of the work can be carried out. Short cuts must be taken, therefore, aiming to make the biggest and easiest savings first followed by finer economies as time permits.

Conclusion

A cumbersome, complicated, untidy system of cost reporting can be most effective if action is taken on its results, yet the smoothest, most straightforward, neatest cost statement can prove useless if not heeded. The techniques discussed in this book can be learned, practised and mastered but at the end of the day it is not the size of the report nor the depth of the calculations that will improve production but the action taken on those figures.

Cost consciousness is an attitude of mind, an attitude that strives to avoid waste in every way from earthmoving plant to office stationery, from fuel consumption to telephone calls, from bricklayers to tea boys, but the methods devised to save such waste must be geared to the savings that can be made otherwise the systems set up will not be cost effective.

Chapter 17

Work study – work measurement

BS 3138:1979 defines work measurement as 'The application of techniques designed to establish the time for a qualified worker to carry out a task at a defined rate of working' in other words, the theoretical time that it should be taking to do the job.

Time study

The main technique of work measurement is time study which is not simply the timing of an operation but a procedure designed to produce standard times or standard outputs for any job of work no matter who is carrying out the work as individuals vary considerably in their output potential. Time study therefore must be taken through three stages.

(1) Timing – to see how long the job is taking.
(2) Rating – to assess the individual being studied against a norm or standard performance.
(3) Relaxation – to allow for appropriate relaxation time.

Before commencing the study it is essential to gather as much detail about the work as possible and record this as a preamble or study top sheet as suggested in Figure 17.1. This top sheet is an important introduction to all studies and must be produced whatever work measurement technique is being used.

(1) Timing

The timing of a particular task is not simply a matter of noting down the time when that task starts and when it finishes but of detailed recording of the time taken to carry out quite short elements of work within the task.

This timing can be carried out by a normal wristwatch: however calculation of the times taken for each element of work is made much easier by the use of a stop watch, recording not in seconds but in 1/100 th of a minute and being of the fly-back type where the watch hand returns to zero when the button is pressed and then immediately recommences timing.

Figure 17.2 illustrates the recording of these times in minutes and decimals of a minute. BS 3139 defines such times as 'observed times'.

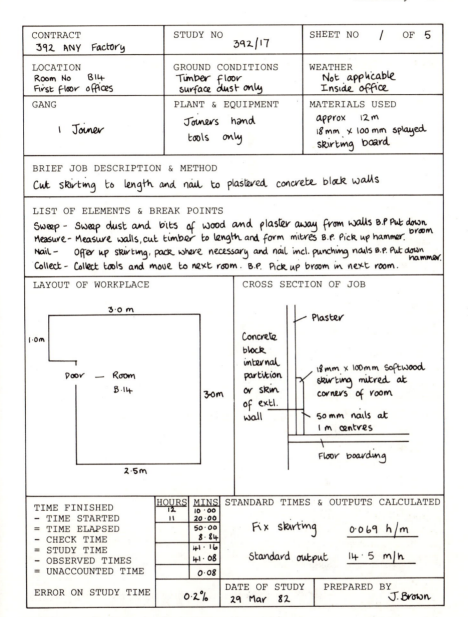

Figure 17.1 Study top sheet showing outline and results of study

Before carrying out a time study it is essential to zero a stop watch to a known clock or wristwatch time and then record the time that elapses before the study starts (TEBS). Similarly at the end of the study the time that elapses after the finish of the study (TEAF) until stop watch and clock time are again synchronised must be noted down. These times are defined in BS 3139 as *check time*. By taking this precaution a check can be made of any error in watch reading that has occurred as shown in Figure 17.1. If this

152 Work study

CONTRACT ANY Factory				STUDY NO 392/17			SHEET NO 2 OF 5	
ELEMENT	Observed time	Rating	Basic time	ELEMENT	Observed time	Rating	Basic time
	Minutes							
Check time	(5·70)		—					
Sweep	1·75	95	1·66					
Measure	4·13	90	3·72					
Sweep	0·65	95	0·62					
Measure	6·92	95	6·57					
Light cigarette	0·41	IT	—					
Measure	4·14	100	4·14					
Nail	12·20	110	13·42					
Sweep	0·38	95	0·36					
Nail	2·62	105	2·75					
Measure	0·71	105	0·75					
Talk	0·22	IT	—					
Nail	5·84	110	6·42					
Blow nose	0·14	IT	—					
Collect	0·97	100	0·97					
Check time	(3·14)		—					
Totals	41·08		41·38					
Check times	(8·84)							

Figure 17.2 Time study continuation sheet

error is more than + or − 2% then the study must be treated with extreme caution and possibly discarded.

(2) Rating

The defined rate of working referred to in the British Standard definition of work measurement is represented in BS 3138:1979 by 'a scale where 0 corresponds to no activity and 100 to standard rating'.

Standard rating

Definition of this 100 standard rating is stated as 'The rating corresponding to the average rate at which qualified workers naturally work, provided that they adhere to the specified method and that they are motivated to apply themselves to their work'. In simple terms this standard rating of 100 can be visualised as a person walking on level ground at a speed of 4 mph; walking at 3 mph would then be at a rating of 75 and 5 mph a rating of 125.

Observed rating

To obtain a true and accurate concept of 100 rating and to be able to observer other ratings to either side of this standard a degree of formal training with an experienced work study engineer is essential. Constant practice and frequent training checks (rating clinics) are necessary to maintain a reliable rating ability. However, it is not a technique to be looked upon in awe as a gift for only the initiated nor is it a con to be ridiculed as being unreliable. The capacity for assessing how well or how badly others are performing is a characteristic inherent in the human race. Work study rating simply sharpens this capacity and provides a numeric scale against which any rate of working can be assessed. The Building Advisory Service of the NFBTE (*the Principles of Incentives for the Construction Industry 1981*, Advisory service for the Building Industry, 18 Mansfield Street, London W1M 9FG), gives as an example the following scale which is to such a broad degree of accuracy that even without training most people could classify their fellow workers into the appropriate category.

Very slow	50
Slow	60
Steady	75
Brisk	100
Fast	120
Very fast	130
Exceptionally fast	150

It follows that with training much finer observed ratings than this can be achieved at first down to the nearest 10 points and with practice to the nearest five.

Although the above table refers to fast and slow workers it must be appreciated that rating is not really a measure of speed but more of effectiveness. BS 3138:1979 defines *to rate* as 'To assess the worker's rate of working relative to the observer's concept of the rate corresponding to standard rating. The observer may take into account, separately or in combination, one or more factors necessary to the carrying out of the task, such as: speed of movement, effort, dexterity, consistency'. It can be seen by this definition that speed is not the only yardstick by which a man's rate of working is assessed, it is in fact, a combination of skill and effort producing a satisfactory result. The ratings of each element of work are recorded as that work is being carried out as *observed ratings* as shown in Figure 17.2.

Elements of relaxation are timed in order to provide an overall check on the watch readings, however, they need not be rated as they are *ineffective time*, i.e. 'That portion of the elapsed time, excluding the check time spent on any activity which is not a specified part of a task', and will therefore be ignored in the calculation of the time for the job.

Basic time

If the rating of an element of work is not at 100 then the observed time for that element has to be corrected in order to bring it to *basic time*, i.e. 'The time for carrying out an element of work or an operation at standard rating'. This process of standardising or normalising the observed times is known as extension and is defined by BS 3138:1979 as 'The process of converting standard time to basic time, i.e. observed time × observed rating/100 (standard rating)', thus an element observed to take five minutes but rated at only 80 rating would take

$$\frac{5 \text{ min} \times 80}{100} = 4 \text{ basic minutes}$$

Contract __ANY Factory__ Study no __392/17__ Sheet no __3__ of __5__

Element:	Sweep									
	75	80	85	90	95	100	105	110	115	120
					1·75					
					0·65					
					0·38					
					2·78					
				=	2·64			Total	2·64	

Element:	Measure									
	75	80	85	90	95	100	105	110	115	120
				4·13	6·92	4·14	0·71			
				4·13	6·92	4·14	0·71			
			=	3·72	=6·57	=4·14	=0·75	Total	15·18	

Figure 17.3 Study extension or normalisation sheet of time study

if the operator had been working at 100 rating. Similarly the same element observed to take only three minutes but rated at 133 rating would take

$$\frac{3 \times 133}{100} = 4 \text{ basic minutes}$$

This extension can be carried out as illustrated in Figure 17.3 on the study sheet. However, where a study lasts for several continuation sheets and keeps repeating the same elements, then an alternative method of collecting the observed times under the appropriate ratings and extending them as a total is much quicker and less likely to lead to mathematical error. The extension is repeated in Figure 17.3 and 17.4 in order to illustrate this alternative.

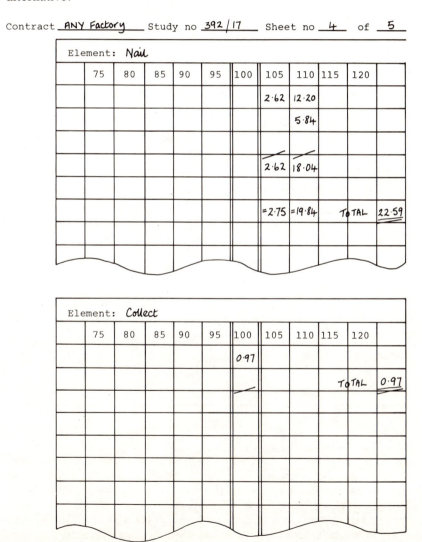

Figure 17.4 Study extension or normalisation sheet of time study

(3) Relaxation

Any elements of relaxation which occur during a time study are to be ignored so that the theoretical time to carry out the task is not affected by the degree of relaxation that any individual person takes. However, an allowance must be made for relaxation as no-one can be expected to work without an occasional stretch, a visit to the toilet or canteen, a smoke or an

```
1. CONSTANT ALLOWANCES:                     E. Air Conditions
                          Men   Women          (excluding climatic factors)
   Personal Needs Allowance 5     7                                      Men   Women
   Basic Fatigue Allowance..  4     4          Well ventilated,or fresh air  0    0
                            ___   ___          Badly ventilated,but no toxic
                             9    11             or injurious fumes........5    5

2. VARIABLE ADDITIONS TO                      Work close to furnaces , etc. 5-15
   BASIC FATIGUE ALLOWANCE                                                  per cent.

   A. Standing Allowance...   2     4
                                               F. Visual Strain
   B. Abnormal Position Allowance              Fairly fine work.........  0    0
   Slightly awkward.........   0     1        Fine or exacting..........  2    2
   Awkward (bending)........   2     3        Very fine or very exacting  5    5
   Very awkward (lying,
        stretching up)......   7     7        G. Aural Strain
                                               Continuous...............  0    0
   C. Weightlifting or                         Intermittent, loud.......  2    2
      Use of Force                             Intermittent, very loud..⎫
      (lifting, pulling or pushing)            High-pitched, loud.......⎬  5    5
                                                                        ⎭
   Weight lifted or force exerted (in kg)
                   2.5........   0    1       H. Mental Strain
                   5..........   1    2       Fairly complex process...  1    1
                   7.5........   2    3       Complex or wide span
                  10 .........   3    4          of attention..........  4    4
                  12.5........   4    6       Very complex.............  8    8
                  15..........   6    9
                  17.5 .......   8   12       I. Monotony: Mental
                  20..........  10   15       Low......................  0    0
                  22.5........  12   18       Medium...................  1    1
                  25..........  14    -       High.....................  4    4
                  30..........  19    -
                  40..........  33    -
                  50..........  58    -

   D. Light Conditions                         J. Monotony: Physical
   Slightly below recommeded                   Rather tedious...........  0    0
        value...............   0     0        Tedious..................  2    1
   Well below..............   2     2        Very tedious.............  5    2
   Quite inadequate........   5     5
```

Figure 17.5 Example of relaxation allowances expressed as percentage on basic times

occasional chat to his fellow workers. Such an allowance is sometimes dealt with by assuming that a worker will only work say 52.5 min in any hour but is more accurate if looked at in the light of the particular task being performed as clearly a welder working inside a 1 m diameter pipe will need more relaxation than say a JCB driver sitting at his controls in the relative comfort of his cab. Many organisations have their own recommendations for these allowances but a widely accepted table is that illustrated by the

International Labour Office (*Introduction to Work Study*, International Labour Office, Geneva, 1959), as shown in Figure 17.5.

The constant allowances of 9% for men and 11% for women are made on all work, the *personal needs allowance* being for such needs as getting a drink, going to the toilet, smoking a cigarette, etc., whilst the *basic fatigue allowance* is to cover the rest or stretching that is required even from a sitting position. The variable allowances cover the degree of difficulty above this constant. Allowances for strain and monotony are rarely required in the construction industry.

In addition a contingency allowance may be necessary to cover small elements of work that did not occur during the study but are known to happen from time to time, e.g. setting out, fuelling up, sharpening tools, etc.

When these relaxation allowances are added to the basic times for the various elements within a job as shown in Figure 17.6 the result is known as the standard time which is the time that it will take to complete a job where a man is working at the standard rating of 100 and takes only that amount of relaxation and contingency time that has been allowed. Where he achieves this over the period of the working day he is said to achieve standard performance which BS 3138:1979 defines as 'the rate of output which qualified workers can achieve without over-exertion as an average over the working day or shift provided they adhere to the specified method and provided they are motivated to apply themselves to their work. This is represented by 100 on the BS scale'.

The standard output of 14.5 m of skirting per hour produced by the time study illustrated is therefore the output that should be produced on that task by the average motivated joiner.

Accuracy of time studies

Of course if the specification or method is different from that described in the study then the standard calculated would not apply. Minor differences such as the cross-section of the skirting, size or quantity of nails, shape or size of the room are not likely to cause a significant change in the standard output. The addition of plugging to brick walls would, however, be significant and would have to be studied separately.

It goes almost without saying that in any mathematically based technique accuracy is of prime importance, however there is a difference between carrying out a work measurement technique accurately and the continual studying of the same task in order to obtain a more accurate result.

Often on a construction site a few hours' study of a single operation is sufficient to give a usable output standard of a sufficient degree of accuracy to enable method improvements to be considered or incentive targets to be set. Days of studying the same operations under slightly different conditions are unlikely to prove a worthwhile investment as work measurement almost always produces results along a line of diminishing returns as shown in the diagram on page 159.

CONTRACT ANY Factory   STUDY NO 392/17   SHEET NO 5 OF 5

JOB DESCRIPTION		BASIC TIME	RELAXATION ALLOWANCES								STANDARD TIME	QUANTITY	UNIT	STANDARD MIN/UNIT	STANDARD H/UNIT	STANDARD OUTPUT
Cut skirting to length and nail to plastered concrete block walls.			CONSTANT	STANDING	ABNORMAL POSITION	USE OF FORCE	CONDITIONS	CONTINGENCIES	TOTAL % ALLOWANCES	RELAXATION TIME						
		MIN	%	%	%	%	%	%	%	MIN	MIN			MIN	HOURS	UNITS PER H
CODE	ELEMENT DESCRIPTION															
SWEEP	Sweep dust and bits of wood and plaster away from walls	2.64	9	2	-	-	-	-	11%	0.29	2.93					
MEASURE	Measure walls, cut timber to length and form mitres	15.18	9	2	1	-	-	-	12%	1.82	17.00					
NAIL	Offer up skirting, pack where necessary and nail including punching nails	22.59	9	2	2	-	-	-	13%	2.94	25.53					
COLLECT	Collect tools and move to next room	0.97	9	2	1	-	-	-	12%	0.12	1.09					
	Total all elements										46.55	11.30	m	4.12	0.069	14.5

Figure 17.6 Calculation of relaxation allowances and summary of time study

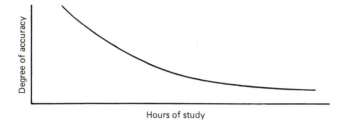

This argument must not be taken however as an excuse to never study the same operation twice. Output standards must be frequently reviewed, different methods, conditions, plant, etc., evaluated but always with the thought in mind that the end must justify the means and that the study man must not be spending his time chasing some statistical degree of accuracy that is anyway not achievable in the environment of a building site.

Having laid down the arguments for relatively short studies providing quick results, there can be no excuse for sloppy study work even if it is carried out under inclement weather conditions. All the principles of carrying out a study in a factory must be followed on a site. The study man must explain what he is going to do, to the foreman in charge and to the men themselves, not necessarily to each individual in a large gang but at least to the leading hands. The stop watch and wristwatch must be properly synchronised to check that at the end the study ties in with real time to at least an accuracy of + or − 2%. The study man should remain standing, not leaning or sitting down, during the period of his observations and take an interest in the reasons for every action or inaction without getting in anyone's way. Study papers should be kept as neat as possible, the study board covered with a transparent plastic cover if weather conditions are unfavourable and as much information as can be obtained about the task, conditions, location, plant, sizes, specifications, etc., written on the study top sheet. The results will then be credible and of use not only in the isolation of the job being studied but also to others and on other sites.

Gang study

The time study example illustrated was for work carried out by one joiner working on his own. Using the same technique it is possible to study two men at the same time provided they are both visible to the studyman and the time of each element is not too short. This can still be done on the same study forms and the basic times added together for each of the men in order to calculate the standard time for the job.

An alternative method which can be used for more than two men is to time the task continuously, i.e. not using the fly back capability of the stop watch which becomes unmanageable as the intervals between elements starting and stopping become proportionally shorter as the number of men being studied increases. As illustrated in Figure 17.7 the watch reading has to be recorded every time any member of the team changes what he is

CONTRACT ANY Factory		STUDY NO 392/18		SHEET NO 2 OF 4		
OBSERVED TIME	ELEMENT	Smith	Jones	Brown	Evans	
00·00	Waiting for concrete	IT	IT	IT	IT	
01·30	Open skip and direct concrete	110				
01·35	Shovel concrete		90	90		
01·53	" "				75	
02·05	Standing				IT	
03·62	Vibrate concrete				80	
04·98	Standing		IT	IT		
04·49	Shovel concrete		85	85		
05·21	Standing	IT				
06·14	Waiting for concrete		IT	IT		
06·70	" "				IT	
08·95	Skip	105				
09·15	Shovel		90	100	90	
10·03	Vibrate				90	
10·14	Stand	IT				
11·09	Shovel	95				
11·20	Skip	100			IT	
12·15	Stand		IT	IT		
12·20	Shovel		100			
12·83	Stand	IT				
13·14	Vibrate				100	
13·25	Shovel			95		
14·12	Wait		IT	IT		

Figure 17.7 Gang study continuation sheet

doing. The rating can then be shown under the appropriate man's name to show which of the men has changed from one element to another or has ceased working. The minutes that each man spends on each element is entered against the appropriate rating column of the study extension sheet for calculation of basic times and subsequently of standard times in the normal way.

Rated activity sampling

A further alternative for timing a number of men, especially useful where not all of the gang are visible from the same position, is a technique known as rated activity sampling. This consists of looking to see what each of the men is doing at frequent regular intervals, usually every one minute. At the same time the man or men being observed are rated so that a time study emerges where every timing is assumed to be of one minute duration.

CONTRACT ANY Factory			STUDY NO 392/19			SHEET NO 2 OF 4					
ELEMENT CODE						DESCRIPTION OF ELEMENT					
S			Hoist and fix scaffold tube and fittings.								
B			Hoist and fix scaffold boards and toe boards.								
IT			Ineffective time								

	Smith			Jones			Brown			Green			
1	90	S	0.90	90	S	0.90	IT			IT			1
2	90	S	0.90	IT			IT			100	S	1.00	2
3	90	S	0.90	90	S	0.90	IT			100	S	1.00	3
4	IT			IT			IT			IT			4
5	IT			IT			IT			IT			5
6	100	S	1.00	100	S	1.00	100	S	1.00	100	S	1.00	6
7	IT			100	S	1.00	100	S	1.00	100	S	1.00	7
8	90	S	0.90	IT			100	S	1.00	100	S	1.00	8
9	95	S	0.95	95	S	0.95	100	S	1.00	100	S	1.00	9
10	IT			IT			100	S	1.00	100	S	1.00	10
11	IT			IT			90	S	0.90	90	S	0.90	11
12	95	B	0.95	95	B	0.95	90	S	0.90	90	S	0.90	12
13	100	B	1.00	95	B	0.95	IT			IT			13
14	100	B	1.00	100	B	1.00	IT			90	S	0.90	14
15	110	S	1.10	100	B	1.00	95	S	0.95	95	S	0.95	15
16	110	S	1.10	95	B	0.95	100	S	1.00	IT			16
17	100	S	1.00	100	B	1.00	100	S	1.00	IT			17
18	100	B	1.00	100	B	1.00	95	S	0.95	IT			18
19	90	B	0.90	90	B	0.90	95	S	0.95	100	B	1.00	19
20	90	B	0.90	90	B	0.90	IT			100	B	1.00	20
21	90	B	0.90	90	B	0.90	IT			90	B	0.90	21
22	IT			IT			IT			90	B	0.90	22
23	IT			IT			IT			90	B	0.90	23
24	100	S	1.00	100	S	1.00	IT			IT			24
25	100	S	1.00	100	S	1.00	IT			IT			25
26	100	S	1.00	100	S	1.00	95	B	0.95	95	S	0.95	26
27	IT			100	S	1.00	95	B	0.95	95	S	0.95	27
28	100	S	1.00	110	S	1.10	95	B	0.95	IT			28
29	100	S	1.00	110	S	1.10	95	B	0.95	90	B	0.90	29
30	IT			IT			90	B		IT			30

Figure 17.8 Rated activity sample continuation sheet

Usually with this type of study the elements of work cannot be as short as with a time study as elements of less than one minute's duration could be missed altogether and the degree of accuracy is clearly going to be restricted by the one minute observations. However, using broad elements it is possible to study quite large gangs as illustrated in Figure 17.8. As with a conventional time study it is essential to record as much detail about the task as possible, the format of the study top sheet shown in Figure 17.2

CONTRACT **ANY Factory** STUDY NO. **392/19** SHEET NO **3** OF **4**										
ELEMENT: S. Hoist and fix scaffold tube and fittings.											
75	80	85	90	95	100	105	110	115	120	
			⊞	⊞	⊞		‖‖‖‖			
			⊞	‖‖‖	⊞					
			‖		⊞					
					⊞					
					⊞					
					‖‖					
			11·00	8·00	27·00		4·00		Total	50·00
			= 9·90	= 7·60	= 27·00		= 4·40		Total	48·90

ELEMENT: B · Hoist and fix scaffold boards and toe boards											
75	80	85	90	95	100	105	110	115	120	
			⊞	⊞	⊞					
			⊞	‖‖‖	‖‖‖‖					
			‖							
			11·00	8·00	9·00				Total	28·00
			= 9·90	= 7·60	= 9·00				Total	26·50

Figure 17.9 Study extension or normalisation sheet of rated activity sample

CONTRACT ANY Factory   STUDY NO 392/19   SHEET NO 4 OF 4

JOB DESCRIPTION				RELAXATION ALLOWANCES												
		BASIC TIME	CONSTANT	STANDING	ABNORMAL POSITION	USE OF FORCE	CONDITIONS	CONTINGENCIES	TOTAL & ALLOWANCES	RELAXATION TIME	STANDARD TIME	QUANTITY	UNIT	STANDARD MINS/UNIT	STANDARD HOURS/UNIT	STANDARD OUTPUT
		MIN	%	%	%	%	%	%	%	MIN	MIN			MIN	HOURS	UNITS PER H
Hoist and fix scaffolding tubes and fittings and boarding including toe boards																
CODE	ELEMENT DESCRIPTION															
S	Hoist and fix scaffold tube and fittings	48.90	9	2	–	–	–	–	12%	5.87	54.77	40	m	1.37	0.023	43.8
B	Hoist and fix scaffold boards and toe boards	26.50	9	2	–	–	–	–	12%	3.18	29.68	20	m	1.48	0.025	40.4
	Total all elements (measured per metre of tube)										84.45	40	m	2.11	0.035	28.4

Figure 17.10 Calculation of relaxation allowances and summary of rated activity sample

being still suitable for a rated activity sample though no check should be necessary of watch reading accuracy.

Depending on the length of the study basic times can either be extended as shown in Figure 17.8 and then simply added together under element headings or for longer studies the observed ratings can be abstracted on an extension sheet as illustrated in Figure 17.9. After abstracting these figures it is advisable to at least check that the number of observations of each element on the study continuation sheets agrees with the minutes recorded on the extension sheet. In the example given Figure 17.8 shows 50 observations of element S and 28 of element B. Figure 17.9 totals 50 observed minutes of element S and 28 observed minutes of element B.

Once the observed minutes have been extended to basic minutes the standard times can be calculated in the normal time study manner as illustrated in Figure 17.10 which produces a standard output for erection of tubes and fittings of 43.8 m of tube per hour and for boarding of 40.4 m of board per hour. A combined time for erecting tubes and boards is shown as 0.035 h per metre of tube.

Study No.	1	2	3	4	5	6	7	8	9	10	11	12
Cumulative basic minutes	43	108	146	215	320	360	420	540				
Cumulative measure	12	20	30	50	70	80	92	120				
Basic minutes per unit	3.6	5.4	4.9	4.3	4.6	4.5	4.6	4.5				

In order to obtain a reliable standard using rated activity sampling, such a study should be continued for at least a few hours and preferably repeated a number of times until an acceptable average output can be accumulated as indicated in Figure 17.11. Results from activity rating studies will be found to vary more widely than from time studies because the timing interval of 1 min is 100 times longer than the centiminute used for time studies.

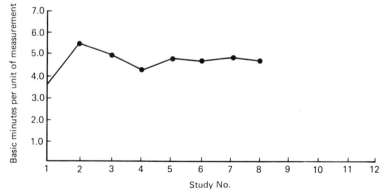

Figure 17.11 Calculation of acceptable basic time by rated activity sample

Synthesis

One of the advantages of breaking a task down into small elements of work with defined break points between the elements is to be able to add elements from one study to elements of another study and so calculate the standard time of a task that has not itself been studied. This technique is known as *synthesis*. In the fixing of a wall shutter for example so many variations to that shutter may be specified each one having an effect on the standard time. The surface of the shutter for instance may be wrot, lined, ribbed or featured. the method of fixing may be by props, clamps or bolts, the surface fixings to the shutter may be unistruts, cardboard tubes, boxes, dovetail blocks, etc. The number of combinations of these variables prevents time studies being carried out on each possibility. However, by adding together the synthetics or element times from various studies the standard time for the required specification can be obtained.

Synthesis is essential in any operation having an element that will vary during the course of the contract as, for instance, in bulk earthmoving where the load and tip elements may remain constant but the haul route will differ depending on the distance of the cut area from the fill area, on the gradient of the haul route, and on the condition of the route.

Element	Time per trip
Load	1.20 basic minutes
Haul 300 m	0.80 ,, ,,
Tip	0.45 ,, ,,
Return 300 m	0.75 ,, ,,
Manoeuvre for loading	0.25 ,, ,,
Total cycle time	3.45 ,, ,,
Relaxation allowance 9%	0.31
	3.76 standard minutes

Figure 17.12 Earthmoving standard calculated by synthesis

Taking as an example a CAT 657 motor scraper loading from a cutting and depositing its load 300 m away in an embankment. The standard time per load could be built up synthetically as illustrated in Figure 17.12 at an average payload of 20 m^3 per load (measured before bulking) this is a standard time of 0.188 min/m^3 or 379 m^3/h.

Analytical estimating

Where the synthetics to produce a standard time for a task are not all known it may be necessary to assess or estimate certain elements. This technique is known as *analytical estimating*. A typical illustration of this could be taken from the previous example if the times of one or more of the elements were not known. For example it does not require a prolonged time study to calculate standard times for loading, tipping and manoeuvring of a scraper but the time taken to haul and return requires much more studywork as so many variables such as distance, gradient, obstructions, conditions occur during the study. As illustrated in Figure 17.13, it may be necessary to assess the time required to carry out these two elements based on knowledge and experience of similar types of work or it may be necessary to base the times on the manufacturer's performance handbook or technical literature, i.e. at 20 m³/load a standard output of 330 m³/h.

Element	Time per trip
Load	1.20 basic minutes
Haul 300 m @ say 25 km/h	0.72 ,, ,,
Tip	0.45 ,, ,,
Return 300 m @ say 25 km/h	0.72 ,, ,,
Manoeuvre for loading	0.25 ,, ,,
Total cycle time	3.34 ,, ,,
Relaxation allowance 9%	0.30
	3.64 standard minutes

Figure 17.13 Earthmoving standard calculated by analytical estimating

Comparative estimating

Where a number of similar, yet not identical, tasks are to be performed it is not necessary to carry out a time study on every different task even if the standard times of the unstudied tasks cannot be built up by synthetics. By taking a series of studied operations (bench marks) a time for each of the unstudied operations can be assessed by comparison with the bench mark jobs. This technique is known as comparative estimating and is particularly useful in studying, for example, the variety of ironmongery in an office complex or for assessing the inumerable one-off jobs that occur on every site which does not warrant study but is not quite the same as anything already timed.

Slotting

An extension of comparative estimating is to use a series of time bands which can either be statistically set up or simply presented as for instance

A	1 min jobs
B	5 min jobs
C	15 min jobs
D	1 h jobs
E	3 h jobs
F	1 day jobs

Every task can then be slotted into one of these categories with the help of a few bench mark jobs that have actually been time studied. This technique is known as slotting.

Predetermined motion time systems (PMTS)

PMTS is a series of work measurement techniques based on the calculation of movements and mental activities, for example a turn of the hand, a movement of the eye, etc. The duration of the average task, the degree of repetition, the vaguaries of site conditions and the changes in methods being used do not usually lend themselves in the construction industry to fine degrees of measurement required by PMTS systems.

Methods – Time Measurement (MTM) forms the basis of most PMTS systems with motions being measured in time measurement units (tmu) of 0.00001 h, i.e. 0.036 s. MTM data is usually expressed at a rating level of MTM 100 which is said to be equivalent to 83* on the BS scale of 0 to 100.

Various levels or grades of measurement exist, the most detailed being MTM–1 which evaluates such basic motions as

RELEASE, REACH, GRASP, MOVE, POSITION

To calculate the standard time of a task by this means can take as long as 150 times the duration of the task so is clearly of little benefit on a construction site. At higher levels MTM–2 combines certain motions, for example the RELEASE, REACH, GRASP of level 1 becomes GET in level 2 and the MOVE, POSITION motions become PUT.

Simplifying even further, MTM–3 combines all of these five basic motions into one motion of HANDLE. Various manufacturing companies and management consultants have simplified MTM even further taking various motions together to form a complete movement or operation that is commonplace within their particular field or branch of industry. Many of these newer levels of MTM use magnetic tape for recording the movement of workmen and microcomputers for analysing the data. Undoubtedly these higher level systems will continue to develop in the future and will become more acceptable to site construction work.

Random activity sampling

At the other end of the scale to PMTS a simple system for evaluating in broad terms the time spent on a contract in carrying out certain activities or lack of activity is a technique known as Random Activity Sampling. For instance if it is required to know how much of joiners' time on a contract is spent fetching and carrying materials, then a sample could be taken by going around the site at irregular intervals and noting down how many joiners were working and how many were fetching materials. If at the end of the day the joiners were seen to be working 150 times and fetching material 50 times, then it can be assumed that during the period of the study the joiners were fetching materials roughly 25% of the time.

It is important when carrying out a sample of this nature to select a period that is typical of the work being carried out and to select random times that do not fall into any particular pattern. Random number tables can be used to set up the random times but this is not usually necessary as long as commonsense is used; for example if out of half a dozen samples per day three of them are taken after tea and meal breaks then a distorted picture will almost certainly result. The number of samples required can be checked statistically, though again commonsense should dictate when sufficient data have been collected. To check the number of observations

---

* Work Measurement by Dennis A Whitmore BSc, MTech, CEng, MIEE, MIERE, MBIM published on behalf of the Institute of Management Services.

required the following formula can be applied with a 95% level of confidence

$$n = \frac{4p(100-p)}{L^2}$$

where n = number of observations required, p = percentage of time spent on a particular activity and L = Limit of error required, i.e. + or − $L\%$. The formula can also be turned around to give the degree of accuracy that is being achieved as follows

$$L = 2\sqrt{\frac{p(100-p)}{n}}$$

Figure 17.14 illustrates a simple activity sample on a site looking at all workmen to see whether they are working or not working. The result can be demonstrated as a pie chart as shown in Figure 17.15.

CONTRACT ANY Factory		STUDY NO 392/20		SHEET NO 2 OF 2			
Description of work All labour on site					Date 15 April 82		
SAMPLE NO.	TIME	WORKING			NOT WORKING		
1	8·05	ℍℍ	ℍℍ	|||	ℍℍ	||	
2	8·13	ℍℍ	ℍℍ	||	ℍℍ	|||	
3	8·22	ℍℍ	ℍℍ	ℍℍ	ℍℍ		
4	8·38	ℍℍ	ℍℍ	|||	ℍℍ	||	
5	8·55	ℍℍ	ℍℍ	||||	ℍℍ			
6	9·06	ℍℍ			ℍℍ	ℍℍ	ℍℍ
7	9·09	ℍℍ	ℍℍ	ℍℍ	ℍℍ		
8	9·19	ℍℍ	ℍℍ	||	ℍℍ	|||	
9	9·30	ℍℍ	ℍℍ	||	ℍℍ	|||	
10	9·38	ℍℍ	ℍℍ	ℍℍ	||||			
11							
12							
13							
14							
15							
16							

Figure 17.14 Random activity sample continuation sheet

Where a work measurement technique has produced a standard time for a task, an activity sample can be used to break down that time into elements as, for instance, in a breakdown of fixing reinforcement, if the time spent in tieing, threading, lacing, etc., was required, or into the type of labour carrying out the elements such as a breakdown of driving pre-cast concrete piles if the time spent by fitters, drivers, hammermen, labourers, etc. was required.

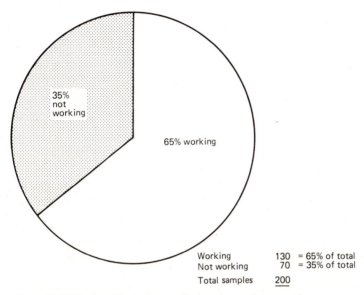

Figure 17.15 Pie chart illustrating result of random activity sample

Random Activity Sampling is normally used without attempting to rate the observed workers and therefore assumes a constant rate of working by the whole workforce. This assumption is of course not true but as the technique is only designed as a spot check or sample it is sufficiently accurate to prompt more detailed study of the locations or tasks that indicate poor productivity or overmanning.

Exercises

(1) An activity sample is carried out to find the amount of time spent transporting materials on site. A pilot study showed that about 30% of hours were spent on transporting materials. Calculate how many observations are needed to be sure of this figure within plus or minus 10%.
(2) Recalculate Exercise 1 to an accuracy of plus or minus 5%.
(3) An activity sample was taken on a concrete mixer to determine how often it was standing. It was found to be standing on 20 observations out of 100, i.e. 20% of the time. Calculate to what degree of accuracy this 20 can be taken.

Chapter 18

Work study – method study

BS 3138:1979 defines *method study* as 'The systematic recording and critical examination of ways of doing things in order to make improvements'. The act of recording work measurement data frequently exposes ideas for method improvement. It may be that such an improvement is so clearly worthwhile that it is immediately accepted by all concerned and adopted without opposition. It is more likely, however, that the benefits from any improvement are less clear cut and may have to be sold hard to stand any chance of adoption. The suggested improvement may require additional capital investment, an increase or reduction in manpower or in plant, changes in authority, encroachment into someone else's empire or removal of their sacred cows or any number of changes in the status quo that will generate objections, reasons, excuses or simply flat refusals to adopt the proposal. Or it may be that no ideas for improvement have shown themselves. In any of these situations the need for detailed analysis of the existing method and the use of method study techniques is necessary to expose, evaluate and sell the potential improvements.

Questioning technique

A method study technique that runs through all other techniques of method improvement is the act of critically examining a task from various viewpoints to see if any stage of that task could be improved or even eliminated.

The questioning is broken down into five fields of enquiry.

(1) The *purpose* for which the task is done.
(2) The *place* at which the task is done.
(3) The *sequence* in which the task is done.
(4) The *person* by whom the task is done.
(5) The *means* by which the task is done.

Primary questions

The primary questions to each of these fields of enquiry establish what the situation is at present.

(1) *What* is achieved?
(2) *Where* is it done?
(3) *When* is it done?
(4) *Who* does it?
(5) *How* is it done?

Each of these statements are then subjected to the question *why* to discover the reason for the present situation.

(1) *Why* is it necessary?
(2) *Why* there?
(3) *Why* then?
(4) *Why* that person?
(5) *Why* is it done that way?

Secondary questions

The secondary questions then examine the alternatives available.

(1) *What* else could be done?
(2) *Where* else could it be done?
(3) *When else* could it be done?
(4) *Who else* could do it?
(5) *How else* could it be done?

Finally a decision is made on the best option to take with the following questions.

(1) *What should* be done?
(2) *Where should* it be done?
(3) *When should* it be done?
(4) *Who should* do it?
(5) *How should* it be done?

As with all work study techniques, the end must justify the means in selecting tasks to be studied but clearly the more work is subjected to this critical examination then the more wasted effort on site will be eliminated.

It is preferable to adopt a written form of the questioning technique as illustrated in Figure 18.1 in order to stimulate ideas from others involved in the method improvement. The series of questions, however, soon becomes an attitude of mind and must at the very least be thought through as a means of approach to all method study problems.

Multiple activity charts

Many tasks on a construction site are carried out by a team of workmen and machines working as a gang. Wherever a task involves the use of men working together or working with machines there is always a need to

OPERATION: Cleaning out saw cuts in concrete slab for mastic filling

	What is achieved?	Why is it necessary?	What else could be done?	What should be done?
PURPOSE	Contraction joint saw cuts cleaned out for mastic filling by sub-contractor	Joints fill up with sludge and debris from site	1. Temporary strip could be inserted immediately after sawing 2. Additional concrete saws could be hired to keep up with mastic sub-contractor.	1. Temporary plastic strip shaped + for easy removal and adequate cover to joint to stop rain and surface water

	Where is it done?	Why there?	Where else could it be done?	Where should it be done?
PLACE	On concrete floor slab	Joints are in slab	Must be at joints	1. At joints 2. Temporary plastic strips to be specially purchased.

	When is it done?	Why then?	When else could it be done?	When should it be done?
SEQUENCE	Immediately before sub-contractor fills joints with mastic	Saw cutting has to be within 24 hrs of placing concrete. Quantity of joints only warrants one visit per week by mastic sub-contractor	Joints would not get contaminated if done up to 24 hrs before mastic filling	1. Strip inserted immediately after sawing 2. Withdrawal immediately before mastic filling 3. Strips collected and cleaned for reuse.

	Who does it?	Why that person?	Who else could do it?	Who should do it?
PERSON	2 Labourers	1. Not skilled operation 2. One man holds up mastic sub-contractor when cleaning out is difficult.	1. Young labourer 2. Sub-contractors labour	1. Saw cutter to insert strips. 2. Sub-contractor to withdraw strips 3. Young labourer to collect and clean strips when other work not available

	How is it done?	Why is it done that way?	How else could it be done?	How should it be done?
MEANS	By raking along joints with narrow metal stick and brushing away from area of joint	No other tool will fit into joint	1. Temporary strip could be pulled out leaving joint clean. 2. Cut could be resawn 3. Design special tool.	Job eliminated in favour of protective strip

Figure 18.1 Critical examination sheet

achieve the optimum balance of these resources, e.g. the right number of labourers placing concrete to the capacity of the delivery vehicles, the right number of lorries carting away spoil to the excavator digging it, the right number of drainlayers laying pipes to the excavator forming the trench.

This can best be demonstrated by the technique known as *multiple activity charts* where each activity involved in the task is plotted on a bar line to a time scale as illustrated in Figure 18.2 which shows an operation of concreting a first-floor slab by concrete skip hoisted from ground level by crane. The banksman fills a $1\,m^3$ skip from the concrete delivery vehicle and two labourers discharge the skip and vibrate the concrete at first-floor level. The whole cycle takes eight minutes. The shaded areas highlight the period of time that each man is forced to be ineffective because there is nothing he can do. This enforced ineffective time is defined by BS 3138: 1979 as *unoccupied time* and the proportion of the *standard cycle time* that he can be working is defined as the *load factor*.

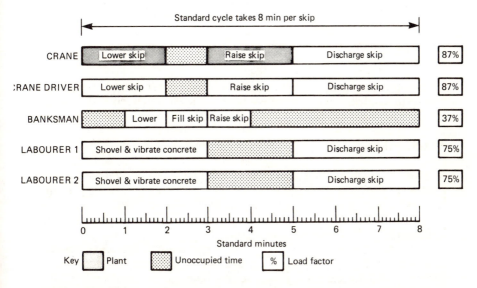

Figure 18.2 Multiple activity chart. Placing concrete – showing existing method proposed

The questioning technique can now be used in an attempt to eliminate as much as possible of this unoccupied time and thereby suggest method improvement which will increase output and/or reduce costs. Figure 18.3 suggests a possible alternative method using two concrete skips. The banksman can now have a full skip ready to be raised as soon as the empty skip is lowered so that waiting for the skip to be filled with concrete is eliminated. The standard cycle now lasts only seven minutes to place $1\,m^3$ of concrete resulting in an improvement in output of 12.5% and a reduction in cost of 11.5% as calculated in Figure 18.4. Using this method comparison sheet a number of alternatives can be considered, some of which may produce improved output but at a higher unit cost, other

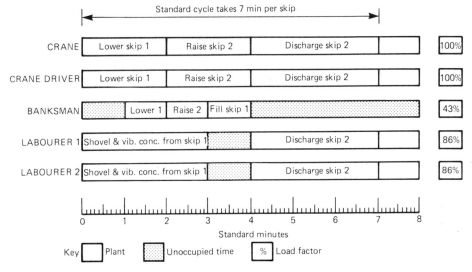

Figure 18.3 Multiple activity chart. Placing concrete – showing proposed method

suggestions may produce lower costs but at a poorer output. Laid out in this format a decision can more easily be made on which method to adopt.

Standard cost

It must be realised that the labour and plant cost at standard output calculated in Figure 18.4 is not the actual cost of placing the concrete but is the cost assuming all work is carried out at standard 100 performance. The actual cost will vary from day to day and no doubt throughout the day depending on the actual performance of the gang at any particular time. If for instance the gang using the existing method of placing concrete calculated in Figure 18.4 were, on average, working at 120 performance then the actual cycle time would be less than the standard cycle time of eight minutes which is based on 100 performance. The cycle time would be

$$8 \text{ min} \times \frac{100 \text{ performance}}{120 \text{ performance}} = \underline{6.67 \text{ min/skip}}$$

the actual output 60 min ÷ 6.67 min/skip = 9 skips/gang h and the actual cost £37.30/h ÷ 9 skips/h = $\underline{£4.14/1 \text{ m}^3 \text{ skip}}$.

The danger is that this actual cost of £4.14 for the existing method will be compared with the proposed method No. 1 at a standard cost of £4.39 and will be immediately rejected. It is essential therefore that all cost comparisons including the existing method are evaluated at the same level of performance, i.e. 100.

Should it be known that certain elements within the cycle are going to be consistently worked at non-standard performance then it is still preferable to decide on the best method based on standard times first and then to

UNIT OF MEASUREMENT per one cubic metre skip	BUILD UP OF GANG COST PER HOUR	STANDARD CYCLE TIME	STANDARD OUTPUT/ GANG H	LABOUR & PLANT COST AT STANDARD OUTPUT	EFFECT ON OUTPUT	EFFECT ON COST
EXISTING METHOD Crane and banksman at ground level. Two labourers at first floor level. One skip	Crane 20·00 Driver 5·00 Banksman 4·00 Labourers 2 @ £4·00 8·00 Skip 0·30 £37·30	8 min	7½ skips	£4−97		
ALTERNATIVE METHOD NO. 1 As existing but two skips	as above 37·30 additional skip ·30 £37·60	7 min	8½ skips	£4−39	+ 12½%	− 11½%
ALTERNATIVE METHOD NO. 2						
ALTERNATIVE METHOD NO. 3						
ALTERNATIVE METHOD NO. 4						
ALTERNATIVE METHOD NO. 5						

Figure 18.4 Method comparison sheet – placing concrete

176 *Work study – method study*

calculate the effect on outputs and costs of the non-standard working, e.g. if part of the work is to be carried out by an apprentice or by someone who consistently produces exceptionally high or low performances. To attempt to build such non-standard working into the calculations in the first place will only lead to confusion.

Relaxation time

Because the concept of 100 performance assumes that the correct amount of relaxation time will be taken so it is right to base multiple activity charts on standard times and not on basic times. Figure 18.5 shows a multiple activity chart for a gang breaking out and carting away tarmac and

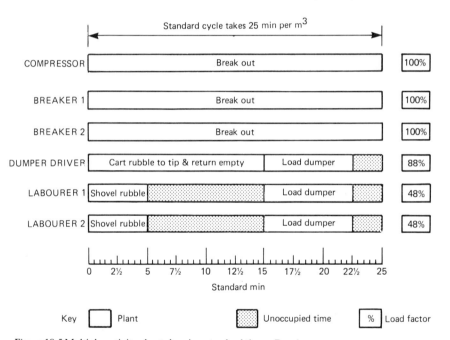

Figure 18.5 Multiple activity chart showing standard times. Break out tarmac and roadstone – showing existing method

roadstone. It would appear, at first glance, that a possible improvement would be to remove labourer No. 2. Certainly the shovelling of rubble ready for loading could be done by labourer No. 1 whilst the dumper is away at the tip. However, there is insufficient unoccupied time, whilst the dumper is present, for labourer No. 1 and the dumper driver to do the loading currently carried out by labourer No. 2. This would seem a pity as labourer No. 2 is left with only 2.5 standard minutes of work in each 25 min cycle whilst labourer No. 1 is only 80% occupied as shown in Figure 18.6.

In such a marginal situation it may be worth considering a breakdown of the standard time as relaxation can clearly be taken during unoccupied

Relaxation time 177

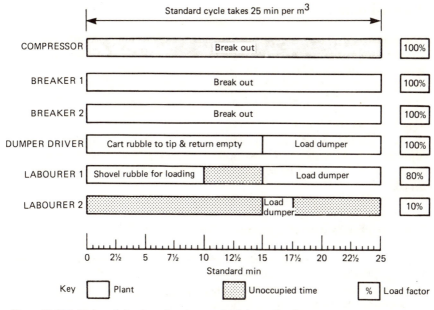

Figure 18.6 Multiple activity chart showing standard times. Break out tarmac and roadstone – showing alternative method

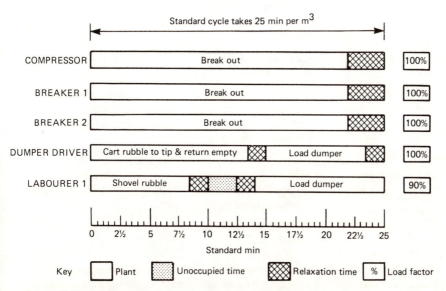

Figure 18.7 Multiple activity chart separating relaxation time and basic time. Break out tarmac and roadstone – showing alternative method

time thus creating a degree of overlap between some of the bar lines. Figure 18.7 illustrates this by taking advantage of this overlap. The remaining work left for labourer No. 2 can now be carried out by labourer No. 1 without increasing the standard cycle time.

Allowed time

The unoccupied time remaining when all possible method improvements have been exhausted cannot be ignored in the setting of incentive scheme targets or in the planning of production capabilities and manpower requirements. An unoccupied time allowance must be added to the appropriate standard times for the tasks but only when carried out by the method as calculated. The resultant time is no longer a standard time because it has been enhanced by the unoccupied time allowance. BS 3138 defines this enhanced time as *allowed time* which may also include such allowances as bonus increments, policy allowances and learner allowances, all of which must never be considered as being part of the standard time for a task.

Process charts and diagrams

Process charts and diagrams are formed by the use of symbols to indicate what happens to a particular item of work, material, piece of plant or element of labour at each point during its progress through a site. BS 3138: 1979 defines these symbols as follows

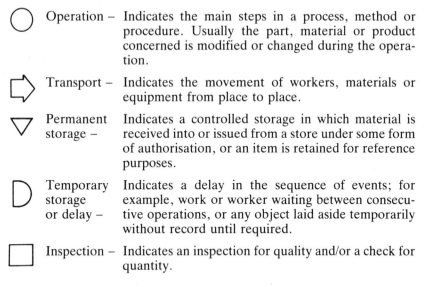

Although of less use in the construction industry than in manufacturing processes, the use of these symbols can assist in highlighting where processing problems such as material handling are likely to occur. By covering a site layout with tracing paper, a number of alternate locations for site offices, stores, etc. and a number of alternate routes for material handling, transport of plant or progression of labour can be considered. The use of these symbols, if necessary in differing colours, can enable a number of variations to be plotted on the same tracing helping to highlight optimum routes, alternative storage needs, possible delays, etc., that may not otherwise become apparent until construction has commenced when resiting of stores, repositioning of gates, weighbridges, checkers' offices, etc., may no longer be economical or even possible.

The layout of production line processes such as precasting yards, concrete mixer set-up, gravel barrow pits, reinforcement cutting and bending yards and large-scale fitters' or joiners' shops need careful consideration of flow from one process to another within that set up. The sequence of events laid out either on a diagram or tabulated as a list can assist in the visualising of the process in order to apply the questioning technique to each event and thus initiate ideas for improvement.

String diagrams

Of similar use is the technique of superimposing on a model or drawing, by the use of string or crayon the route that a process is to take. This can be particularly useful where excessive material handling is likely to be involved on a site to and from different locations. After inserting nails or pins at key locations on a site layout journeys between these locations can be represented by rubber bands or string loops, each thread representing a fixed number of journeys. After all known journeys have been plotted in this way any areas likely to be congested will be apparent by the thickness of the thread. Specification for temporary roads can be adjusted depending on the degree of traffic indicated, and other possible means of access, craneage, construction progress, method, etc., can all be considered where little use or overuse is likely to be made of a particular access area. Flow chart symbols for major tasks can also be superimposed on such a string diagram to further highlight congestion and possible delays to the work. Again the questioning technique can be applied at the highlighted locations in order to exhaust all possible alternatives.

Photographic techniques

Photographic method improvement techniques require specialist equipment and are not often used in the construction industry. Briefly they can be described as follows

Cycle graph. A constant light source is attached to the item under study and is traced on a photograph thus forming a continuous line of light in a similar manner to the string in a string diagram.

Chronocycle graph. Similar to a cycle graph but the light source is switched on and off at known intervals thus forming a line of pear shaped spots, the pointed end indicating the direction of movement and the spacing indicating the speed of movement thus highlighting congestion or delays in a process or material handling situations.

Micromotion photography. Film taken by movie camera at faster than normal speed for later projection in slow motion. This enables closer study of a task than the eye can normally capture.

180 Work study – method study

Memomotion photography. Film taken by movie camera at slower than normal speed for later projection showing a speeding up of the activity. This is a form of visual activity sampling and is useful for exposing for instance the number of journeys that have to be made to the stores, the number of round trips in the period of film or the degree of side tasks being undertaken which are not part of the main task.

Exercises

(1) The standard output for loading of excavated material to a lorry by JCB 806 has been calculated as 40 m^3/h and the hourly output of Foden 16 tonne tipper lorries to transport the material to tip and return as 15 m^3/h. A decision therefore has to be made on whether to employ only two lorries and have unoccupied time on the excavation or employ three lorries and have unoccupied time on the transport.

Using the method comparison sheet illustrated in Figure 18.4 and current running costs for plant and drivers, calculate the relative merits of the two options.

(2) You have carried out time studies on the operation of excavate drain trench by JCB 3 lay and butt joint 150 mm diameter porous concrete pipes and backfill with stone direct from delivery wagons. The cross-section of the trench is 500 mm wide at top and bottom × 1 m deep.

Standard times from your studies are as follows

A Excavate by JCB 3 and deposit at back of trench	0.050 h/m^3
B Hold traveller for boning in trench	0.005 h/m
C Sight profile for boning in trench	0.005 h/m
D Hand trim bottom of excavation	0.030 h/m^2
E Lay and butt joint pipes	0.040 h/m
F Banksman to lorries backing up to trench	0.010 h/m^3
G Spread and level stone by hand at top of trench	0.060 h/m^2

The present method is for the JCB 3 to excavate followed by a ganger and two labourers who bone in and trim the bottom of the trench. No banksman to the JCB is thought necessary. Two pipelayers then lay the pipes. A further ganger and two labourers supervise the tipping and hand trimming of the stone.

(1) Plot the operations on a multiple activity chart, per 100 m of trench.
(2) Suggest and plot an improved method that could be adopted as a short-term measure.
(3) Plot the effect of using a JCB 806 excavator as a long-term improvement. Standard time assumed with same width bucket = 0.020 h/m^3.
(4) Calculate the relative costs and outputs of the different methods based on current costs for the plant and labour.

Chapter 19
TYMLOG recorder

A problem of producing multiple activity charts lies in the number of time studies that have to be carried out in order to calculate the standard times for the various interrelated tasks. It is always more difficult to properly time and rate two men carrying out an operation unless they work in unison, which rarely happens, and when a gang grows to three, four or more men, normal time study and even rated activity sampling techniques become, to say the least, difficult. The solution can be to break the operation down into various tasks and study each task separately but the time required for such an exercise is usually beyond the resources of most construction sites.

A number of systems exist which rely on data collected by tape recorders and later analysed manually or by computer to provide standard times. These systems do, however, require specialist training. An instrument at present on the market known as the TYMLOG* enables a study to be made on 14 men or 14 tasks at the same time. Briefly, it consists of a rotating drum around which pressure-sensitive paper is fed. Any of 14 buttons can be pressed ON making contact with the carbonised paper and thus creating a line as the drum revolves (see Figure 19.2). Each of the buttons can be assigned to either a man in the gang or to a certain task or a combination of both. For counting cycles or deliveries a button can be depressed for about 20 s, just long enough to make a mark, whilst other lines can, at the same time, show the degree of activity taking place on the tasks selected. The number of buttons available is in most cases quite sufficient though it must be borne in mind that if each button represents a different task then the numbers of buttons for each task must be the maximum number of men that may be carrying out that task at any one time. Thus, taking an extreme example, four men carrying out three tasks would require 12 buttons if all four worked together on each of the three tasks. In practice this is not usually the case though a few spare buttons need to be reserved for the sudden purge by more than the usual numbers of men on a particular element being timed. For the odd helping hand from outside the gang, a written note of the contribution and its timing can be made from a conventional stop watch or wristwatch.

---

* Available from Nantglyn Engineering Products Limited, Ridgewood, Colgate, Horsham, Sussex RH12 4TA.

182 TYMLOG recorder

As there are no facilities for rating with this type of recorder a blanket rating will need to be assessed in order to convert observed times to basic time (unless 100 rating is assumed throughout). This can easily be done by noting down at appropriate intervals the ratings observed and then averaging these ratings for each task at the end of the study. The resultant print-out from this machine is an instant multiple activity chart of observed times, the number of minutes in each bar line being easily counted for subsequent calculation of standard times and consideration of method improvements as illustrated by the following example.

Example 19.1 Method improvement – drainage

The land drains to a site are constructed to the cross-section shown in Figure 19.1. A study of the work using a TYMLOG recorder produces the chart shown in Figure 19.2, the work content of each task being as defined in Figure 19.3. A summary of the number of men involved and blanket

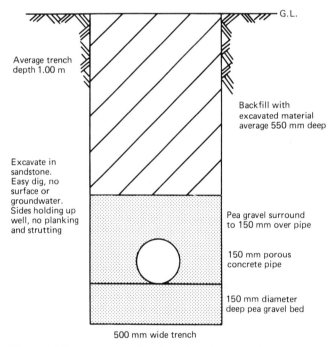

Figure 19.1 Typical cross-section through drainage trench

ratings for each task are shown in Figure 19.4. Average costs per hour are laid down in Figure 19.5. The length of trench completed during the study was 100 m. Quantities for each task can therefore be calculated as shown in Figure 19.6.

Example (1) Using blank TYMLOG or other suitably drawn paper as a multiple activity chart plot possible method improvements and compare their relative costs and outputs on a method comparison sheet.

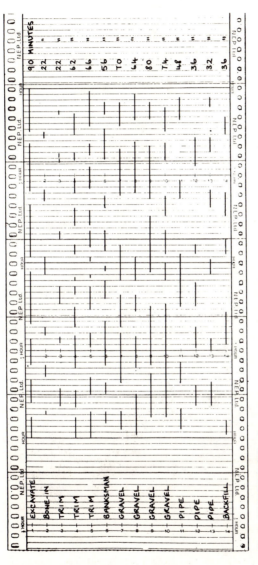

Figure 19.2 TYMLOG print-out

184 TYMLOG recorder

Title	Button number	Definition of work content
Excavate	1	JCB 3 wheeled excavator digs trench, partly deposits spoil at side of trench for later backfilling and partly loads to lorries for carting away to tip (lorries not studied) JCB stands while ganger bones-in levels
Bone-in	2	Ganger sights along profiles and gives instructions to labourers on levels
Trim	3–5	Two labourers and sometimes ganger trim bottom of trench with shovel and throw spoil out to side of trench. Depth of trim about 50 mm average. Negligible trimming to sides of trenches. One labourer holds traveller while ganger sights along profile to check levels
Banksman	6	Banksman guides lorry delivering gravel from local supplier to edge of trench, opens tail board and signals to driver when and how much to tip, manoeuvring the lorry along the trench until empty. Also sights along profiles and gives instructions to labourers on levels of gravel to base of trench and top of pipe surround. This operation sometimes carried out by labourer from gravel levelling gang
Gravel	7–10	The four labourers shovel the gravel to the correct levels, one of them holding the traveller when the banksman sights in the levels. Alternatively another of the labourers sights in along profiles if banksman is working elsewhere in trench. Ganger and labourers move between work on base and surround as available
Pipe	11–13	Pipes have been delivered and run out adjacent to the top of the trench where one or more of the three pipelayers roll pipes to edge of trench, pass or drop them down to where they lay them including setting up string line and levelling with spirit level. Pipes are butt jointed against each other, no other jointing or pointing involved
Fill	14	CAT 931 Traxcavator loading shovel loads spoil from side of trench and tips over gravel then uses bucket as a blade to doze ground to level. Tends to go backwards and forwards over same ground, therefore rated at only 80 rating

Figure 19.3 Definition of work content in each task

Task	Number of men	Totals of observed times	Blanket ratings observed	Basic times calculated
Excavate	1 driver	90 min	100	90 min
Bone-in	1 ganger	22 min	100	22 min
Trim	2 labourers	150 min	90	135 min
Banksman	1 banksman	60 min	110	66 min
Gravel	4 labourers	288 min	80	230 min
Pipe	3 pipelayers	116 min	90	104 min
Backfill	1 driver	36 min	80	29 min

Figure 19.4 Summary of times and ratings during study

Type of plant or labour	Cost per hour
JCB 3 including fuel, spares and maintenance	£10.00
CAT 931 including fuel, spares and maintenance	£14.00
Drivers, gangers and pipelayers	£ 5.00
Banksmen and labourers	£ 4.00

Figure 19.5 Average costs per hour

Task	Measurements	Quantity	Unit of measurement
Excavate	100 m x 0.5 m x 1.0 m	50	m^3
Trim including bone-in	100 m x 0.5 m	50	m^2
Gravel including B/M	2/100 m x 0.5 m	100	m^2
Pipe	100 m	100	m
Backfill	100 m x 0.5 m x 0.55 m	27½	m^3

Figure 19.6 Quantities of work completed during study

Example (2) Using the proposed method which you choose as the most acceptable calculate the allowed time (including unoccupied time) for the whole operation expressed in man h/m of trench.

Example (3) Using the quantities as measured in Figure 19.6 calculate the standard times (exluding unoccupied time) for each of the individual tasks per unit of measurement.

Example (1) Comparison of methods

It can be seen immediately from the chart in Figure 19.2 that a considerable amount of ineffective time is occurring and, as frequently happens during a study a number of improvements will already have been brought to mind either by the study man, by site management or by the gang itself. The vertical ruling on the chart is at two-minute intervals which makes calculation of the line lengths relatively easy and sufficiently accurate for this kind of study. These times have been entered as observed times at the end of the chart in Figure 19.2 and totalled against each task in Figure 19.4. Basic time can then be calculated by the following formula.

$$\text{Basic time} = \text{Observed time} \times \frac{\text{observed rating}}{\text{standard rating, i.e. 100}}$$

From this point the calculations are treated as a conventional time study. Figure 19.7 shows the addition of relaxation allowances and Figure 19.8 shows the existing method plotted as a multiple activity chart with all times converted to basic times and relaxation allowances shown as dotted lines. The ineffective time now shows even more clearly and the cycle time of the whole operation can be seen to be governed by the JCB, i.e. 90 basic min

JOB DESCRIPTION		RELAXATION ALLOWANCES								PER 100 m
Excavate drain, lay 150mm dia. porous concrete pipe, bed and surround with pea gravel and backfill	BASIC TIME	CONSTANT	STANDING	ABNORMAL POSITION	USE OF FORCE	CONDITIONS	CONTINGENCIES	TOTAL % ALLOWANCES	RELAXATION	STANDARD TIME
ELEMENT DESCRIPTION	MIN	%	%	%	%	%	%	%	MIN	MIN
Excavate by JCB 3	90	9	–	–	–	–	–	9%	8	98
Bone-in	22	9	2	1	–	–	–	12%	3	25
Trim	135	9	2	2	1	–	–	14%	19	154
Banksman	66	9	2	1	–	–	–	12%	8	74
Gravel	230	9	2	2	2	–	–	15%	35	265
Pipe	104	9	2	2	1	–	–	14%	15	119
Backfill by CAT 931	29	9	–	–	–	–	–	9%	3	32
NON STUDIED WORK (Times assessed by comparative estimating)										
Excavate by JCB 806 say	60	9	–	–	–	–	–	9%	5	65
Backfill by D4 dozer say	40	9	–	–	–	–	–	9%	4	44
Backfill by JCB 3 say	35	9	–	–	–	–	–	9%	3	38
Banksman excluding boning in gravel say	33	9	2	1	–	–	–	12%	4	27

Figure 19.7

spent excavating plus a further 22 min of unoccupied time when the ganger is boning in and the JCB has to stop working. Relaxation allowance of the driver can be assumed to be taken during his unoccupied time giving a total cycle of 112 standard min/100 m of trench.

Having established the cycle time both the standard output per gang hour and the standard cost per 100 m of trench can be calculated (as shown

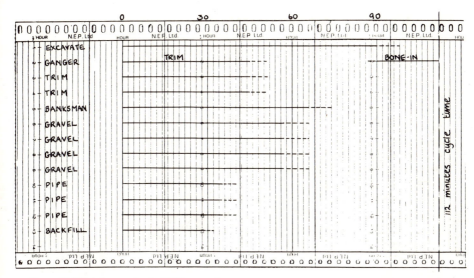

Figure 19.8 Existing method – 13 No. men in gang

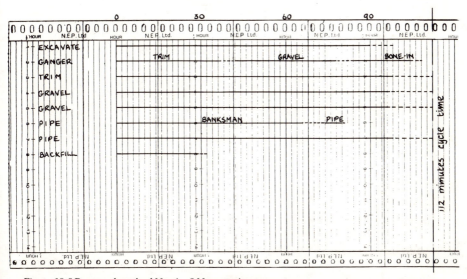

Figure 19.9 Proposed method No. 1 – 8 No. men in gang

in Figure 19.14). Here the cost per hour of the gang including plant costs, fuel and plant consumables has been built up. The cycle time is taken from the multiple activity chart in Figure 19.8 and the standard output calculated thus

$$\text{Standard output} = \text{Unit of measurement} \times \frac{1 \text{ hour}}{\text{cycle time}}$$

$$= 100 \text{ m} \times \frac{60 \text{ min}}{112 \text{ min}}$$

$$= 53.57 \text{ m/h}$$

The labour and plant cost at standard output is calculated as follows

$$\text{Standard cost} = \text{Gang cost/h} \times \frac{\text{cycle time}}{1\,\text{h}}$$

$$= £82.00 \times \frac{112\,\text{min}}{60\,\text{min}}$$

$$= £153.07$$

The standard cost is not to be confused with the actual cost as explained on page 174. The effect on output and cost of any proposed method improvements can now be expressed by comparison with the standard output and standard cost produced by the existing method.

The first and most immediate action that could be taken in such a situation is to reduce the number of men in the gang. From Figure 19.7 it can be seen that the work content time in 100 m of trench totals 767 min. Dividing this by the cycle time of 112 min for the existing method the minimum number of men that could carry out the work, assuming all were working at 100 performance, would be $767 \div 112 = 7$ men. However, the practicalities of, say, expecting the CAT driver to jump out of his machine and assist with the levelling of gravel further up the trench mean that this minimum may not be possible. In fact the proposed method No. 1 illustrated in Figure 19.9 suggests a team of eight men as being the minimum practical figure at least until the other more long-term changes can be made. The effects of this suggestion on cost per 100 m can be seen in Figure 19.14.

The next suggestion illustrated in Figure 19.10 is to use a less expensive machine for backfilling as the CAT 931 appears too powerful for this work and is therefore only about 30% effective. Any loss in output by a smaller machine is unlikely to change the overall cycle time therefore the effect on cost (as illustrated in Figure 19.14) is advantageous without loss in overall output.

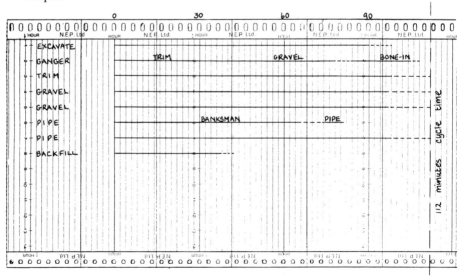

Figure 19.10 Proposed method No. 2 – 8 No. men in gang. Backfill by D4

Of prime importance in any attempt at method improvement is the reduction in the overall cycle time. Proposed method No. 3 illustrated in Figure 19.11 therefore looks at the effect of using a more powerful machine for excavating and in the absence of any study on the proposed machine an assessment is made in Figure 19.7 of the output by JCB 806 compared with that of the studied JCB 3. The result is a decrease in cycle time with an appropriate reduction in cost and increase in output as calculated in Figure 19.14.

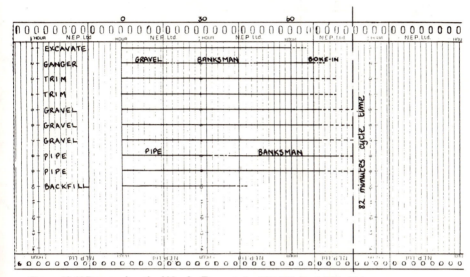

Figure 19.11 Proposed method No. 3 – Excavate by JCB 806. Backfill by D4. 10 No. men in gang

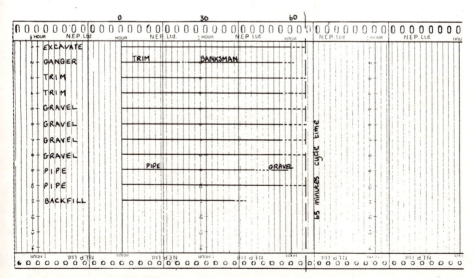

Figure 19.12 Proposed method No. 4 – Excavate by JCB 806. Backfill by D4. Level by laser. 11 No. men in gang

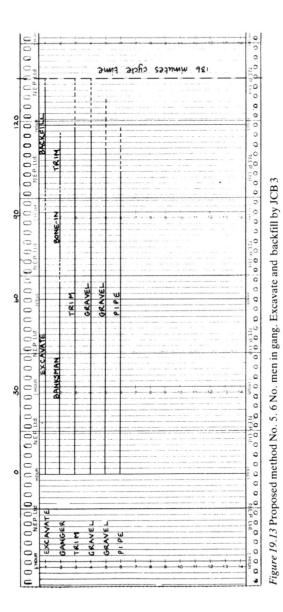

Figure 19.13 Proposed method No. 5. 6 No. men in gang. Excavate and backfill by JCB 3

Proposed method No. 4 illustrated in Figure 19.12 looks at the problem of the JCB having to stop whilst levelling of the trench is carried out by profiles and travellers. A laser could relieve the ganger of this duty and also relieve the banksman when levelling in the gravel. Again the cycle time decreases with an appropriate reduction in cost and increase in output as shown in Figure 19.14.

An alternative method of improvement may be to allow the JCB 3 excavator, which can also double as a loading shovel, to carry out the backfilling. Proposed method No. 5 illustrated in Figure 19.13 shows that using this method the cycle time will increase, thus producing a lower output. This may be totally unacceptable for reasons of progress and regardless of any savings in cost. However, as Figure 19.14 indicates, although output against the existing method would fall by almost 20% the effect on cost of this proposal would be a saving of 45%: furthermore this change in method could be quickly implemented because it does not involve additional plant. Thus with a variety of proposals prepared, a management decision can be taken on which method to adopt. Such factors as continuity of the drainage trench, progress required, availability of the work, plant requirements elsewhere, in-company plant available, particularly favourable hire rates from outside plant hire companies, etc., will need to be considered but with the calculations shown in Figure 19.14 acting as the foundation for such consideration.

Once implemented a further study of the method selection may expose yet more suggestions for improvement. An important aspect of this kind of work is that much of the labourer's tasks cannot be seen to be finalised. That is, they are a matter of degree. The degree of trimming required to the trench bottom, the degree of levelling required to the gravel bed and to the gravel surround. Having operated the gang with too many men the chances are that the required degree of trimming and levelling has been exceeded, yet this is what has been timed. It is asking a lot of the study man to expect this aspect to be included to satisfactory accuracy in his blanket ratings. Studies therefore need to be continued using even less labourers than those suggested in order to determine where the minimum degree of trimming and levelling truly lies. Comparisons need to be made with similar operations in other studies and from other sites. In short the method improvement is never complete, there is always scope for looking again and yet again until such time as the cost of additional studies fails to justify the improvements that can be made. In the construction industry the operation is frequently completed before this point is reached.

Example (2) Allowed time for selected method

Because allowed time must include unoccupied time (BS 3138) the allowed time for the selected method can be calculated as follows

Allowed time = standard cycle time × number of men

e.g. if proposed method No. 5 were selected then

Allowed time = 136 min × 6 men
             = 13.6 man hours/100 m

UNIT OF MEASUREMENT per 100 m of trench	BUILD UP OF GANG COST per hour	STANDARD CYCLE TIME	STANDARD OUTPUT/ GANG H	LABOUR & PLANT COST AT STANDARD OUTPUT	EFFECT ON OUTPUT	EFFECT ON COST
EXISTING METHOD Excavate by JCB 3 Backfill by CAT 931 13 No men	JCB 3 10-00 CAT 931 14-00 6 men @ £5-00 30-00 7 men @ £4-00 28-00 £82-00	112 min	53½ m	£153-07		
ALTERNATIVE METHOD NO. 1 Excavate by JCB 3 Backfill by CAT 931 8 No men	JCB 3 10-00 CAT 931 14-00 5 men @ £5-00 25-00 3 men @ £4-00 15-00 £61-00	112 min	53½ m	£113-87	NIL	-26%
ALTERNATIVE METHOD NO. 2 Excavate by JCB 3 Backfill by D4 dozer 8 No men	JCB 3 10-00 D4 dozer say 9-00 5 men @ £5-00 25-00 3 men @ £4-00 12-00 £56-00	112 min	53½ m	£104-53	NIL	-32%
ALTERNATIVE METHOD NO. 3 Excavate by JCB 806 Backfill by D4 dozer 10 No men	JCB 806 say 15-00 D4 dozer say 9-00 5 men @ £5-00 25-00 5 men @ £4-00 20-00 £69-00	82 min	73 m	£94-30	+36%	-38%
ALTERNATIVE METHOD NO. 4 Excavate by JCB 806 Backfill by D4 dozer level by laser 11 No men	JCB 806 say 15-00 D4 dozer say 9-00 Laser say 2-00 5 men @ £5-00 25-00 6 men @ £4-00 24-00 £75-00	65 min	92 m	£81-25	+72%	-47%
ALTERNATIVE METHOD NO. 5 Excavate by JCB 3 Backfill by JCB 3 6 No men	JCB 3 10-00 3 men @ £5-00 15-00 3 men @ £4-00 12-00 £37-00	136 min	44 m	£83-87	-18%	-45%

Figure 19.14 Method comparison sheet

If other work is available for any of the men during their unoccupied time then due allowance must be made for this in calculating the allowed time. For example, if alternative work were possible for the CAT 931 during 50% of its time then proposed method No. 1 could be further reduced in cost by 50% of the loader and driver rates giving a gang cost of £51.50/h and a standard cost of £93.13/100 m. If this method were then selected the allowed time would be

Allowed time = 112 min × 7.5 men
             = 14.0 man hours/100 m

Without this alternative work for the loader the allowed time for proposed method No. 1 would be

Allowed time = 112 min × 8 men
             = 14.9 man hours/100 m

These examples illustrate that an allowed time is only correct for a stated method.

Example (3) Standard time of tasks

Although, as illustrated above, the allowed time will vary from method to method, the same cannot be said for the standard time. Provided the plant

JOB DESCRIPTION	STANDARD TIME	QUANTITY	UNIT	STANDARD MINS/UNIT	STANDARD HOURS/UNIT	STANDARD OUTPUT
Excavate drain, lay 150 mm dia. porous concrete pipe bed and surround with pea gravel and backfill						
ELEMENT DESCRIPTION	MINS			MINS	HOURS	UNITS PER H
Excavate by J.C B 3	98	50	m³	1·96	0·033 h/m³	30 m³/h
Bone in	179	50	m³	3·58	0·060 h/m²	17 m²/h
Trim						
Banksman						
Gravel	339	100	m²	3·39	0·057 h/m²	18 m²/h
Pipe	119	100	m	1·19	0·020 h/m	50 m/h
Backfill by CAT 931	32	27½	m³	1·16	0·019 h/m³	53 m³/h

Figure 19.15

used and the nature of the work remain as studied then the standard time will remain static regardless of any unoccupied time. Figure 19.7 shows the calculation of standard times per 100 m of trench which is the unit of measurement most suitable for producing the multiple activity charts and method comparisons. Figure 19.15 now illustrates these times in more conventional units of measurement for use elsewhere, e.g. different sizes of trench, different pipes, backfill, etc. Without a specific method these standard times can only be added together on the assumption that no unoccupied time exists – an assumption that, unfortunately, is rarely true when men and machines are dependent upon one another as can be seen from Figures 19.9 to 19.13 which indicate a degree of unoccupied time in every method proposed.

Exercises

(1) Plot multiple activity charts or other possible method improvements of the land drain example and compare their relative costs and outputs with the existing method.

(2) Calculate the standard (including unoccupied time) for the existing method and each of the proposed methods shown in the TYMLOG exercise.

(3) A similar trench has to be constructed but using 300 mm diameter porous concrete pipe in a trench 1.50 m deep and 600 mm wide. If the standard output for laying these 300 mm diameter pipes is 20 m/man hour, plot multiple activity charts of various methods that could be used and compare their relative costs and outputs with those of the existing gang.

Chapter 20

Bulk earthmoving – scrapers or dump trucks

The choice of the correct team of bulk earthmoving equipment is of paramount importance to the profitability of any contract where large quantities of excavation and/or fill are involved. A saving of only a few pence per m^3 on a contract requiring several weeks of an earthmoving team can result in profitability being increased by tens of thousands of pounds. It is therefore essential that costs of different methods and team sizes are fully compared before work is commenced and monitored at least weekly throughout the period of the work. There are many variables to be considered as the following calculations will illustrate. However, in the absence of company records, for these variables much assistance can be obtained from manufacturers' literature.

It must first be recognised that although a driver of a machine may be on site for say 60 h/week the machine itself may only work for, say, 75% of this time, i.e. 45 hours/week. The main factors governing this percentage being the age of the machine, its usage and degree of maintenance. With the current cost of fuel it is relevant to differentiate between these as attendance hours and working hours as illustrated in the build-up of costs per machine calculated in Figure 20.1. A number of different choices of team will then need to be developed as shown in Figure 20.2 in order to subsequently test for the most economical team practicable.

The likely output of each scraper or dump truck team must also be calculated. This will depend on

(1) the payload, i.e. the measured cubic metres carried per load. This will vary with the type of material being dug
(2) the haul speed. This will vary with the gradient and condition of the haul route
(3) the return speed which will depend also on the gradient and condition of the haul route
(4) the time taken to load and to tip. With scrapers this will depend on both gradients and on the type of material. With dump trucks the key factor is the machine being used for loading into the dump truck.

The payload can be checked by counting the number of round trips of the scrapers or dump trucks and comparing this with a measured quantity

196 Bulk earthmoving – scrapers or dump trucks

BUILD UP OF COST PER MACHINE CAT 627 SCRAPER	COST PER WORKING H
Weekly hire £ 1000 ÷ 45 working hours per week	£ 22 : 22
Spares & consumables £ 100 per week ÷ 45 working hours per week	£ 2 : 22
Fuel 100 litres per working hour @ 15 p per litre	£ 15 : 00
Driver 60 attendance hours @ £ 5:00 per hour	
= £ 300 per week ÷ 45 working hours per week	£ 6 : 67
TOTAL COST PER WORKING HOUR	£ 46 : 11

BUILD UP OF COST PER MACHINE D 8 DOZER	COST PER WORKING H
Weekly hire £ 700 ÷ 45 working hours per week	£ 15 : 56
Spares & consumables £ 100 per week ÷ 45 working hours per week	£ 2 : 22
Fuel 50 litres per working hour @ 15 p per litre	£ 7 : 50
Driver 60 attendance hours @ £ 5:00 per hour	
= £ 300 per week ÷ 45 working hours per week	£ 6 : 67
TOTAL COST PER WORKING HOUR	£ 31 : 95

BUILD UP OF COST PER MACHINE CAT 16 GRADER	COST PER WORKING H
Weekly hire £ 800 ÷ 45 working hours per week	£ 17 : 78
Spares & consumables £ 250 per week ÷ 45 working hours per week	£ 5 : 56
Fuel 40 litres per working hour @ 15 p per litre	£ 6 : 00
Driver 60 attendance hours @ £ 5:00 per hour	
= £ 300 per week ÷ 45 working hours per week	£ 6 : 67
TOTAL COST PER WORKING HOUR	£ 36 : 00

Figure 20.1 Build-up of costs of machines for earthmoving by scraper

for the same period. The times can be obtained by time study or other work study techniques.

Figure 20.3 assumes the following

(1) payload 11 m^3/load
(2) haul speed 350 m/standard min
(3) return speed 350 m/standard min
(4) load and tip 2.5 standard min/load

The choice of numbers of scrapers per team will depend largely on the loading time and the average haul length. Too many scrapers on a short haul will clearly cause congestion and would be better utilised as two independent teams. Too few scrapers on a long haul will under utilise the

BUILD UP OF TEAM COST PER WORKING HOUR				
TEAM CHOICE (k) __2__			TEAM CHOICE (k) __3__	
__2__ No of __CAT 627__ @ £ __46·11__	£ 92:22	__3__ No of __CAT 627__ @ £ __46·11__	£ 138:33
__2__ No of __D8__ @ £ __31·95__	£ 63:89	__2__ No of __D8__ @ £ __31·95__	£ 63:89
__1__ No of __Grader__ @ £ __36·00__	£ 36:00	__1__ No of __Grader__ @ £ __36·00__	£ 36:00
Bowsers, slashers, fitters, etc. £__500__ ÷ __45__h	£ 11:11	Bowsers, slashers, fitters, etc. £__600__ ÷ __45__h	£ 13:33
TEAM COST (m)	£ 203:22	TEAM COST (m)	£ 251:56
TEAM CHOICE (k) __4__		TEAM CHOICE (k) __5__	
__4__ No of __CAT 627__ @ £ __46·11__	£ 184:44	__5__ No of __CAT 627__ @ £ __46·11__	£ 230:56
__2__ No of __D8__ @ £ __31·95__	£ 63:89	__3__ No of __D8__ @ £ __31·95__	£ 95:83
__1__ No of __Grader__ @ £ __36·00__	£ 36:00	__1__ No of __Grader__ @ £ __36·00__	£ 36:00
Bowsers, slashers, fitters, etc. £__600__ ÷ __45__h	£ 13:33	Bowsers, slashers, fitters, etc. £__700__ ÷ __45__h	£ 15:56
TEAM COST (m)	£ 297:67	TEAM COST (m)	£ 377:94

BUILD UP OF TEAM COST PER WORKING HOUR			
TEAM CHOICE (k) __6__		TEAM CHOICE (k) __7__	
__6__ No of __CAT 627__ @ £ __46·11__	£ 276:67	__7__ No of __CAT 627__ @ £ __46·11__	£ 322:78
__3__ No of __D8__ @ £ __31·95__	£ 95:83	__3__ No of __D8__ @ £ __31·95__	£ 95:83
__1__ No of __Grader__ @ £ __36·00__	£ 36:00	__1__ No of __Grader__ @ £ __36·00__	£ 36:00
Bowsers, slashers, fitters, etc. £__700__ ÷ __45__h	£ 15:56	Bowsers, slashers, fitters, etc. £__800__ ÷ __45__h	£ 17:78
TEAM COST (m)	£ 424:06	TEAM COST (m)	£ 472:39
TEAM CHOICE (k) __8__		TEAM CHOICE (k) ____	
__8__ No of __CAT 627__ @ £ __46·11__	£ 368:89	___ No of _____ @ £ ___	£ :
__3__ No of __D8__ @ £ __31·95__	£ 95:83	___ No of _____ @ £ ___	£ :
__1__ No of __Grader__ @ £ __36·00__	£ 36:00	___ No of _____ @ £ ___	£ :
Bowsers, slashers, fitters, etc. £__800__ ÷ __45__h	£ 17:78	Bowsers, slashers, fitters, etc. £___ ÷ ___h	£ :
TEAM COST (m)	£ 518:50	TEAM COST (m)	£ :

Figure 20.2 Build-up of various team costs for earthmoving by scraper

e	f = e ÷ b HAUL TIME	g = e ÷ c RETURN TIME	h = d+f+g TOTAL CYCLE TIME	i = 60÷h LOADS/H	j = a x i M³/H OF ONE MACHINE	k TEAM CHOICE	l = j x k M³/H OF TEAM	m TEAM COST	n = m ÷ l COST/M³
500	1·43 1·43	1·43 1·43	5·36 5·36	11·20 11·20	123 123	2 3	246 370	£203·22 £251·56	82p 68p
750	2·14 2·14	2·14 2·14	6·79 6·79	8·84 8·84	97 97	3 4	292 389	£251·56 £297·67	86p 77p
1000	2·86 2·86	2·86 2·86	8·21 8·21	7·30 7·30	80 80	3 4	241 321	£251·56 £297·67	£1·04 93p
1250	3·57 3·57	3·57 3·57	9·64 9·64	6·22 6·22	68 68	4 5	274 342	£297·67 £377·94	£1·09 £1·10
1500	4·29 4·29	4·29 4·29	11·07 11·07	5·42 5·42	60 60	4 5	238 298	£297·67 £377·94	£1·25 £1·27
1750	5·00 5·00	5·00 5·00	12·50 12·50	4·80 4·80	53 53	5 6	264 317	£377·94 £424·06	£1·43 £1·34
2000	5·71 5·71	5·71 5·71	13·93 13·93	4·31 4·31	47 47	5 6	237 284	£377·94 £424·06	£1·60 £1·49
2250	6·43 6·43	6·43 6·43	15·36 15·36	3·91 3·91	43 43	6 7	258 301	£424·06 £472·39	£1·64 £1·57
2500	7·14 7·14	7·14 7·14	16·79 16·79	3·57 3·57	39 39	6 7	236 275	£424·06 £472·39	£1·80 £1·72
2750	7·86 7·86	7·86 7·86	18·21 18·21	3·29 3·29	36 36	7 8	254 290	£472·39 £518·50	£1·86 £1·79
3000	8·57 8·57	8·57 8·57	19·64 19·64	3·05 3·05	34 34	7 8	235 269	£472·39 £518·50	£2·01 £1·93

Figure 20.3 Cost per m³ for earthmoving by scraper

BUILD UP OF COST PER MACHINE CAT 769 DUMP TRUCK	COST PER WORKING H
Weekly hire £__900__ ÷ __45__ working hours per week	£ 20 : 00
Spares & consumables £ __250__ per week ÷__45__ working hours per week	£ 5 : 56
Fuel __40__ litres per working hour @ __15__ p per litre	£ 6 : 00
Driver __60__ attendance hours @ £__5:00__ per hour	
= £__300__ per week ÷__45__ working hours per week	£ 6 : 67
TOTAL COST PER WORKING HOUR	£ 38 : 22

BUILD UP OF COST PER MACHINE POCLAIN HC 300 EXCAVATOR	COST PER WORKING H
Weekly hire £__900__ ÷ __45__ working hours per week	£ 20 : 00
Spares & consumables £ __250__ per week ÷ ____ working hours per week	£ 3 : 33
Fuel __40__ litres per working hour @ __15__ p per litre	£ 6 : 00
Driver __60__ attendance hours @ £__5:00__ per hour	
= £ __300__ per week ÷__45__ working hours per week	£ 6 : 67
TOTAL COST PER WORKING HOUR	£ 36 : 00

Figure 20.4 Build-up of costs of machines for earthmoving by dump truck

BUILD UP OF TEAM COST PER WORKING HOUR

TEAM CHOICE (k) __2__		TEAM CHOICE (k) __3__	
__2__ No of __CAT 769__ @ £ __38·22__	£ 76:44	__3__ No of __CAT 769__ @ £ __38·22__	£ 114:67
__1__ No of __HC 300__ @ £ __36·00__	£ 36:00	__1__ No of __HC 300__ @ £ __36·00__	£ 36:00
__1__ No of __D8__ @ £ __31·94__	£ 31:94	__1__ No of __D8__ @ £ __31·94__	£ 31:94
__1__ No of __Grader__ @ £ __36·00__	36·00	__1__ No of __Grader__ @ £ __36·00__	36·00
Bowsers, slashers, fitters, etc. £__500__ ÷ __45__ h	£ 11:11	Bowsers, slashers, fitters, etc. £__600__ ÷ __45__ h	£ 13:33
TEAM COST (m)	£ 191:50	TEAM COST (m)	£ 231:94

TEAM CHOICE (k) __4__		TEAM CHOICE (k) __5__	
__4__ No of __CAT 769__ @ £ __38·22__	£ 152:89	__5__ No of __CAT 769__ @ £ __38·22__	£ 191:11
__1__ No of __HC 300__ @ £ __36·00__	£ 36:00	__1__ No of __HC 300__ @ £ __36·00__	£ 36:00
__1__ No of __D8__ @ £ __31·94__	£ 31:94	__1__ No of __D8__ @ £ __31·94__	£ 31:94
__1__ No of __Grader__ @ £ __36·00__	36·00	__1__ No of __Grader__ @ £ __36·00__	36·00
Bowsers, slashers, fitters, etc.£__600__ ÷ __45__ h	£ 13:33	Bowsers, slashers, fitters, etc. £__700__ ÷ __45__ h	£ 15:56
TEAM COST (m)	£ 270:17	TEAM COST (m)	£ 310:61

Figure 20.5 Build-up of various team costs for earthmoving by dump trucks

e	$f = e \div b$ HAUL TIME	$g = e \div c$ RETURN TIME	$h = d+f+g$ TOTAL CYCLE TIME	$i = 60 \div h$ LOADS/H	$j = a \times i$ M³/H OF ONE MACHINE	k TEAM CHOICE	$l = j \times k$ M³/H OF TEAM	m TEAM COST	$n = m \div l$ COST/M³
500	1.43 1.43	1.43 1.43	9.86 9.86	6.09 6.09	97 97	2 3	195 200*	£191.50 £231.94	£1-16 £1-16
750	2.14 2.14	2.14 2.14	11.29 11.29	6.32 5.32	85 85	2 3	170 200*	£191.50 £231.94	£1-13 £1-16
1000	2.86 2.86	2.86 2.86	12.71 12.71	4.72 4.72	76 76	2 3	151 200*	£191.50 £231.94	£1-27 £1-16
1250	3.57 3.57	3.57 3.57	14.14 14.14	4.24 4.24	68 68	2 3	136 200*	£191.50 £231.94	£1-41 £1-16
1500	4.29 4.29	4.29 4.29	15.57 15.57	3.85 3.85	62 62	3 4	185 200*	£231.94 £270.17	£1-25 £1-35
1750	5.00 5.00	5.00 5.00	17.00 17.00	3.53 3.53	56 56	3 4	169 200*	£231.94 £270.17	£1-37 £1-35
2000	5.71 5.71	5.71 5.71	18.43 18.43	3.26 3.26	52 52	3 4	156 200*	£231.94 £270.17	£1-48 £1-35
2250	6.43 6.43	6.43 6.43	19.86 19.86	3.02 3.02	48 48	3 4	145 193	£231.94 £270.17	£1-60 £1-40
2500	7.14 7.14	7.14 7.14	21.29 21.29	2.82 2.82	45 45	4 5	180 200*	£270.17 £310.61	£1-50 £1-55
2750	7.86 7.86	7.86 7.86	22.71 22.71	2.64 2.64	42 42	4 5	169 200*	£270.17 £310.61	£1-60 £1-55
3000	8.57 8.57	8.57 8.57	24.14 24.14	2.49 2.49	40 40	4 5	159 199	£270.17 £310.61	£1-70 £1-56

* MAXIMUM OUTPUT DEPENDENT ON EXCAVATOR

Figure 20.6 Cost per m³ for earthmoving by dump trucks

attendant plant such as dozers, graders, fuel and water bowsers. As a rule-of-thumb the number of scrapers required can be assessed as the total scraper cycle time divided by the load and tip time. Dividing by the load time only does not allow sufficient time for the back up if push loading the scrapers. The number of vehicles required in a dump truck team can be assessed as the output of the loading machine during the time taken for one trip of the dump truck divided by the average payload of the dump truck.

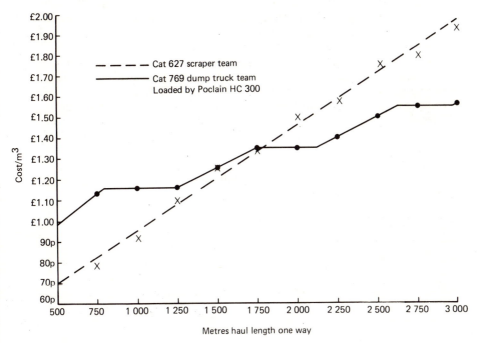

Figure 20.7 Comparison of earthmoving costs at various haul lengths

In the example of a dump truck team in Figures 20.4–20.6 the output of the excavator has been time studied and found to be 200 m³/standard h. The payload and dump truck speeds have been taken as follows

(1) payload 16 m³/load
(2) haul speed 350 m/standard min
(3) return speed 350 m/standard min
(4) load and tip 7 standard min/load

On a 2500 m haul the dump truck cycle time is therefore

(2500 m haul ÷ 350 m/min) + (2500 m return ÷ 350 m/min) + (7 min load and tip)

= 21.29 standard min

In this period the excavator would dig and load

$$200 \text{ m}^3/\text{h} \times \frac{21.29 \text{ min}}{60 \text{ min}} = 70.97 \text{ m}^3$$

202 Bulk earthmoving – scrapers or dump trucks

At a payload of 16 m³/load the number of dump trucks would be

$$\frac{70.97}{16} = 4.44$$

i.e. possibly four or five dump trucks. Figure 20.6 indicates that from a cost viewpoint four is the cheaper but five would produce better progress for the contract.

The resultant costs of the two methods calculated in Figures 20.1–20.3 and 20.4–20.6 can now be plotted graphically as Figure 20.7 in order to determine at what haul length it becomes cheaper to use dump trucks rather than scrapers for the bulk earthmoving operation.

Note. There is a danger in assuming that once this break-even point has been calculated that it will remain about the same for ever more. This of course is not so as a recalculation using a payload of 17 m³ for the dump truck instead of 16 m³ will show when the break-even point will be found to have dropped from around 1800 m to 1400 m. Clearly any changes in hire, fuel or repair costs will also require a revised calculation as will loading time and haul speeds.

The complexity of these variations lend themselves to computer assistance as illustrated in Chapter 30 though this is of course not essential to carry out the calculations.

Chapter 21

Ready-mixed or site-mixed concrete

The use of ready-mixed concrete on a construction site is certainly of assistance in easing the burden of running a contract. Twenty m³ one day, 200 m³ the next, are only a phone call away and alternative sources are always keen to do business should one supplier not provide the required service. Some of the advantages and disadvantages of using ready-mixed concrete are listed in Figure 21.1. In many cases too the cost of purchasing ready-mixed concrete is less than the cost of mixing on site. This is not, however, always so and it is prudent on any contract containing a fair quantity of concrete to obtain quotations for both the supply of ready-mixed concrete and the supply of cement, sand and gravel in order to consider the cost of site mixing.

The following example traces the steps to be taken through such a comparison. Such detailed build-up is not always necessary depending on the size and type of batching plant or mixer required; however, each element of expenditure must be somewhere considered if the comparison is to be realistic.

Material costs are to be the best quotes obtainable after all discounts have been deducted and due consideration given to winning one's own aggregates rather than purchasing from an outside supplier. Having obtained all quotations the first stage is to compare the costs of the two sets of materials because if the combined cost of the cement, sand and aggregate mixed in the appropriate proportion is more expensive than the supply of ready-mixed concrete then no further calculation need be made. If, however, a margin is left below the cost of mixing then that is the budget for site mixing and transporting which must be beaten to make site mixing a worthwhile proposition. The following example illustrates this first step in the comparison

Materials costs/m³

Ready-mixed concrete

Best quotation after discount	£32.00/m³

Site-mixed concrete

300 kg cement @ £50.00 tonne	£15.00	
600 kg sand @ £6.00 tonne	£ 7.20	
1200 kg aggregate @ £5.00 tonne	£ 3.00	£25.20/m³

∴ Budget for mixing and transporting	£ 6.80/m³

204 Ready-mixed or site-mixed concrete

At any time during the period of the contract the purchase price of any of the above materials may change or the design mix of the concrete may be changed. The budget should then be recalculated using the current rates and mixes.

Where more than one major mix of concrete is to be used a mean average can be calculated and differing wastage factors used if checks on

Advantages	Disadvantages
For work at or below ground level, i.e. floors, foundations, kerb races, drain surrounds; when direct access is possible, no site transport is needed	Where the concrete cannot be deposited directly into the formwork, the cost and time of handling must be considered, e.g. if tower crane would hoist concrete direct from mixer
When local aggregates are scarce, difficult arrangements may be by-passed	The careful timing of deliveries required by the need for prior notice, demands closer programming and control of site work with penalties for failure
When labour is scarce, the demand may be reduced	Deliveries may be delayed by plant breakdown, urgent demand by a more important customer, or traffic hold-up
When labour is short on site as at the commencement or completion of a contract a quick start on sub-structure or a finishing item can be made possible	Deliveries must be checked frequently to ensure that the full quantities charged are in fact received. When shortages are discovered, then extra material or credits must be claimed, this can be time absorbing and is often missed
Work that must be completed within a short period, i.e. factory shut down, weekend or holiday; or large units that must be placed continuously, can be done more quickly	When small quantities are required or at the end of the day, part loads are a problem if material is not to be wasted
On restricted sites the working space required for mixer and stock-piles can be avoided	Undefined work or jobs like drain surrounds are difficult to measure for delivery quantities
The facilities for control of quality at a central mixing plant may ensure greater uniformity of mix	Compliance with the specification for mixing must be checked off site. Compression tests for specified strength take time, the only alternative to breaking out unsatisfactory concrete is to claim compensation for the risk of leaving it in position
If concrete requirements are intermittent, then plant standing time is saved	
Most large plants have steam heating equipment so that concrete can still be produced in cold weather	

Figure 21.1 Use of ready-mixed concrete

material wastage prove that the wastage factor is not common to both methods. The comparison for cutting and bending rod reinforcement shown in Figure 22.1 illustrates these two points.

The next step in the comparison is to decide on the size and type of mixer set up required. This will depend on peak outputs necessary during the course of the contract bearing in mind that occasional surges in requirements can be satisfied by supplementing the site-mixed concrete with

FIXED COSTS

Rent of land for mixer set up (taken as fixed cost because usually rented for a fixed period).

	£
1 ha for 1 year @ £5000/ha/year	5000
Survey, legal expenses and rates	1000
Fencing (none required)	Nil
	£6000

Preparation of batching plant area and access roads
Grub up bushes and undergrowth and demolish existing out-house

	£
2 labourers 8 h @ £4.00/h	64
Level area and excavate foundations for batcher	
Excavator/loader 24 h @ £10.00/h	240
Driver 24 h @ £5.00/h	120
2 labourers 40 h @ £4.00/h	320
Pipelayer 16 h @ £5.00/h	80
Concrete foundations	
1 joiner 30 h @ £5.00/h	150
4 labourers 16 h @ £4.00/h	256
60 m^3 × 2 tonne/m^3 hardcore @ £3.00 tonne	360
40 m^3 ready-mixed concrete @ £35 tonne	1400
Shuttering materials, holding down bolts, etc.	250
Drainage materials	150
	£3390

Provision of services

	£
Electricity connection and cut-off charges	900
Water connection and cut off charges	700
	£1600

Transport of batching plant from plant depot

	£
Batching plant 1 No. @ £400	400
Cement silo 1 No. @ £200	200
Excavator/loader 1 No. @ £200	200
	£800

Erect batching plant

	£
Hire of mobile crane 5 days £240.00 day	1200
4 No. fitters for 100 h @ £5.00/h	2000
1 No. electrician for 50 h @ £5.00/h	250
1 No. welder for 50 h @ £5.00/h	250
4 No. labourers for 100 h @ £4.00/h	1600
MS channels and sleepers for aggregates	500
Ground sheets and miscellaneous materials	400
Special provisions for heating aggregates and water	1500
	£7700

Dismantle batching plant

	£
Hire of mobile crane 3 days @ £240.00/day	720
4 No. fitters for 50 h @ £5.00/h	1000
1 No. electrician for 20 h @ £5.00/h	100
4 No. labourers for 50 h @ £4.00/h	800
	£2620

Reinstate area (including recovering topsoil from tip)

	£
Compressor 1 No. @ £400.00/week	400
4 No. labourers 40 h @ £4.00/h	640
Excavator/loader 24 h @ £10.00/h	240
Driver 24 h @ £5.00/h	120
Hired lorry and driver 24 h @ £12.50/h	300
Grade, cultivate and seed by landscape sub-contractor	500
	£2200

	£
Total of fixed costs	**£24 310**

Figure 21.2 Site mixing and transporting of concrete

ready-mixed. Hire charges of different mixers and ease of erection must be considered against these required outputs. The fixed costs of setting up and dismantling the batcher and its associated equipment can now be assessed as illustrated in Figure 21.2.

The next step is to decide on the average number of transporter vehicles required. The output of the transporters can be calculated as the average hours of concreting each week divided by the time taken per round trip in the transporter, times the capacity of the transporter

$$40\,\text{h} \div 35\,\text{min} \times 4.25\,\text{m}^3$$
$$= 291\,\text{m}^3/\text{week}$$

In the example the contract contains $55\,000\,\text{m}^3$ of concrete but of this $50\,000\,\text{m}^3$ is required during a 30-week period, the other 5000 being spread over the remainder of the two-year contract.

It is only worth considering therefore the 30-week period, i.e. $50\,000\,\text{m}^3$ of concrete which will require an average output of $50\,000 \div 30$ weeks $= 1667\,\text{m}^3/\text{week}$ to be delivered from the batcher. This will require

$$1667\,\text{m}^3 \div 291\,\text{m}^3/\text{transporter} = 5.7\,\text{transporters}$$
$$\text{say an average of 6 transporters}$$

VARIABLE COSTS PER WEEK

	£
Batching plant (including fuel, oils and plant consumables)	
Hire of batcher 1 No. @ £2500/week	2500
Hire of cement silo 1 No. @ £120/week	120
Hire of excavator/loader 1 No. @ £400/week	400
Batcher driver 50 h @ £5.00/h	250
Excavator/loader/driver 45 h @ £5.00/h	225
Fitter/electrician/welder 3 × 8 h @ £5.00/h	120
	£3615
Transport concrete (average 6 No. lorry mounted concrete transporters)	
Hire of transporter (including fuel, etc.) 6 No. @ £400.00/week	2400
6 No. drivers 45 h @ £5.00/h	1350
Fitter 16 h @ £5.00/h	80
	£3830
Running costs (not included in cost of plant)	
Electricity charges	150
Water charges	25
	£175
Total variable costs per week	£7620

Figure 21.3 Site mixing and transporting of concrete

The variable costs of running the batcher and delivery fleet can now be assessed as illustrated in Figure 21.3. The weekly running costs of mixing concrete on site should therefore be £7620 which for average output is equivalent to

$$£7620 \div 1667\,\text{m}^3/\text{week} = £4.57/\text{m}^3$$

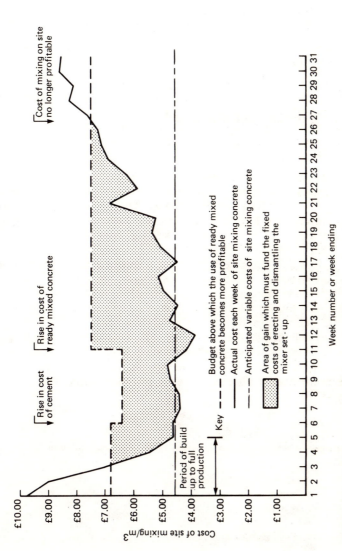

Figure 21.4 Weekly cost of mixing concrete on site

To this must be added the fixed costs of
£24 310 ÷ 50 000 m³   £2.49/m³

giving a total cost of   £5.06/m³
The budget previously calculated is   £6.80/m³

giving a saving if mixed on site of   £1.74/m³

which is a potential saving on the contract of 50 000 m³ @ £1.74

= £87 000

Monitoring the costs of site-mixed concrete

Once the decision has been made to mix concrete on site, procedures must be set up to monitor the actual cost of mixing. In the first few weeks it is unlikely that the site will achieve less than the budgeted limit of £6.80/m³ because it takes time to get gangs laying the concrete into full production. However, once work is in full swing the weekly costs should be around the estimated variable cost of £4.57/m³. As work proceeds this will fluctuate but must be monitored regularly. As soon as concrete production shows signs of falling to the extent where weekly mixing costs climb above the current budget limit then serious thought must be given to changing to ready-mixed concrete as the point will have been reached where the batcher is no longer paying its way. Figure 21.4 illustrates how this can be monitored by plotting the budget, including any changes in that budget against the actual weekly costs. Cumulative costs should not be used because they will include the fixed costs. These fixed costs must be ignored in such a comparison even if they have not yet been fully recovered as there is no point in continuing to mix concrete on site at a cost per m³ that is higher than the cost of purchasing ready-mixed concrete.

Chapter 22
Purchase or site cut and bend reinforcement

The procedure for comparison of costs between the purchase of rod reinforcement in random lengths for cutting and bending on site and the purchase of ready-cut and bent reinforcement labelled and bundled ready for use follows a similar pattern to the previous chapter. The first step is to determine the difference in cost between the best material quotations for the two methods as illustrated in Figure 22.1. Unlike concrete where one mix may dominate the site, rod reinforcement will invariably be in varying quantities of differing diameter with possibly a mixture of mild steel and high tensile reinforcement. An average must therefore first be calculated bearing in mind that more powerful equipment may be needed to cut and bend high tensile reinforcement. Almost certainly a difference in wastage will result from cutting and bending on site which must be taken into

Mild steel rod reinforcement purchased cut and bent

	£
400 tonnes 32 mm @ £240/tonne	96 000
500 tonnes 25 mm @ £235/tonne	117 500
400 tonnes 20 mm @ £235/tonne	94 000
200 tonnes 16 mm @ £240/tonne	48 000
100 tonnes 12 mm @ £270/tonne	27 000
50 tonnes 10 mm @ £280/tonne	14 000
1650 tonnes	£396 500

∴ Average cost = £396 500 ÷ 1650 tonnes = £240.30/tonne

Allow for wastage £240.30/tonne + 2.5% = £246.31/tonne

Mild steel rod reinforcement purchased in random lengths

400 tonnes 32 mm @ £205/tonne	82 000
500 tonnes 25 mm @ £200/tonne	100 000
400 tonnes 20 mm @ £200/tonne	80 000
200 tonnes 16 mm @ £205/tonne	41 000
100 tonnes 12 mm @ £225/tonne	22 500
50 tonnes 10 mm @ £230/tonne	11 500
1650 tonnes	£337 000

∴ Average cost = £337 000 ÷ 1650 tonnes = £204.24/tonne

Allow for wastage £204.24/tonne + 7.5% = £219.56/tonne

∴ Budget for site cutting and bending £26.75/tonne

Figure 22.1 Comparison of material quotations for reinforcement

account in such a comparison and possibly some scrap value allowed for the wasted steel though this can be taken into account in the percentage allowed for wastage.

The fixed costs of setting up the steel cutting and bending yard can be calculated along similar lines to the concrete batching plant though much less resources are likely to be required than for mixing concrete. For the purpose of the example a fixed cost of £1800 is assessed and variable costs as calculated in Figure 22.2 of £1115/week.

Variable costs per week

	£
Cutting and bending yard (including fuel, oils and plant consumables)	
Hire of bar shearing machine 1 No. @ £50.00/week	50
Hire of bar bending machine 1 No. @ £50.00/week	50
3 No. Steelfixers 45 h @ £5.00/h	675
1 No. Labourer 45 h @ £4.00/h	180
1 No. Fitter 1 h @ £5.00/h	5
	960
Transport (including fuel, oils and plant consumables)	
Hire of agricultural tractor 2 days @ £75.00/week	30
Hire of trailers 2 No. @ £20.00/week	40
1 No. tractor driver 16 h @ £5.00/h	80
1 No. Fitter 1 h @ £5.00/h	5
	£155
Running costs (all included above)	Nil
Total variable costs per week	£1115

Figure 22.2 Variable costs per week for site cutting and bending reinforcement

Taking a period of 30 weeks for cutting and bending reinforcement the average weekly output will be

1650 tonnes ÷ 30 weeks = 55 tonnes/week

The weekly running costs of site cutting and bending are therefore £1115 ÷ 55 tonnes/week = £20.27/tonne

To this must be added the fixed costs of

£1800 ÷ 1650 tonnes   £1.09/tonne

giving a total cost of   £21.36/tonne
The budget previously calculated is   £26.75/tonne

giving a saving if cut and bent on site of   £5.39/tonne

which is a potential saving on the contract of 1650 tonnes @ £5.39

                 = £8893

As with the site mixing of concrete it is essential to monitor the weekly cost of site cutting and bending similar to that illustrated in Figure 21.3.

Chapter 23
Hours of work

The hours actually worked on a construction site can have an appreciable effect on the costs of that site. This can come about for a number of reasons

(1) amount of overtime worked
(2) cost of absenteeism
(3) hours effectively at work
(4) shift working
(5) use of labour-only sub-contractors

Amount of overtime worked

Although working overtime attracts premium payments of time and a half or double time, there are other payments such as travelling time and fares allowance which apply by the day, holiday stamps which apply by the week and CITB levy which applies by the year, all of which remain constant regardless of the amount of overtime worked. The premium payments have the effect of increasing costs per man hour whereas the other fixed costs have the effect of reducing the cost per man hour if overtime is worked. The optimum number of hours that a site should work to keep its hourly costs to a minimum will vary according to each of these factors as illustrated in Figures 23.1, 23.2 and 23.3 which show a hypothetical situation on a site for 5, 6 and 7-day weeks. Alteration of any of the factors involved will dramatically change these calculations and it must not be assumed that the results which are plotted graphically in Figure 23.4 will apply to any situation other than the one taken as an example.

Such calculations having so many variables lends itself to computer assistance and would be relatively easy to set up using VISICALC which is discussed in Chapter 30. The following variables must be considered and assumptions made on site averages in order to produce such a calculation. The figures assumed have been summarised where appropriate in Figure 23.5.

Overtime worked	Nil	1 hour/day	2 hours/day	3 hours/day	4 hours/day	5 hours/day	6 hours/day
Hours per week	39	44	49	54	59	64	69
Costs per week							
Basic rate	70.20	79.20	88.20	97.20	106.20	115.20	124.20
Time & a half overtime	–	4.50	9.00	13.50	18.00	18.00	18.00
Double time	–	–	–	–	–	9.00	18.00
Costs not related to hours	45.75	45.75	45.75	45.75	45.75	45.75	45.75
Total subject to GNI	115.95	129.45	142.95	156.45	169.95	187.95	205.95
Graduated National Insurance	15.89	17.73	19.58	21.43	23.28	25.75	27.40
Costs not subject not GNI	20.92	20.92	20.92	20.92	20.92	20.92	20.92
Total cost per week	152.76	168.10	183.45	198.80	214.15	234.62	254.27
Total cost per hour	3.92	3.82	3.74	3.68	3.63	3.67	3.69

Figure 23.1 Payroll costs per man hour for a five-day week

Overtime worked	4 hours Saturday	7 hours Saturday	1 hour/day + 7 h Sat	2 hours/day + 7 h Sat	3 hours/day + 7 h Sat	4 hours/day + 7 h Sat	5 hours/day + 7 h Sat	6 hours/day + 7 h Sat
Hours per week	43	46	51	56	61	66	71	76
Costs per week								
Basic rate	77.40	82.80	91.80	100.80	109.80	118.80	127.80	136.80
Time & a half overtime	3.60	3.60	8.10	12.60	17.10	21.60	21.60	21.60
Double time overtime	–	5.40	5.40	5.40	5.40	5.40	14.40	23.40
Costs not related to hours	53.37	53.37	53.37	53.37	53.37	53.37	53.37	53.37
Total subject to GNI	134.37	145.17	158.67	172.17	185.67	199.17	217.17	235.17
Graduated National Insurance	18.41	19.89	21.74	23.59	25.44	27.29	27.40	27.40
Costs not subject to GNI	22.81	22.81	22.81	22.81	22.81	22.81	22.81	22.81
Total cost per week	175.59	187.87	203.22	218.57	233.92	249.27	267.38	285.38
Total cost per hour	4.08	4.08	3.98	3.90	3.83	3.78	3.77	3.76

Figure 23.2 Payroll costs per man hour for a six-day week

Overtime worked	14 hours Weekend	1 hour/day + 14 h Weekend	2 hours/day + 14 h Weekend	3 hours/day + 14 h Weekend	4 hours/day + 14 h Weekend	5 hours/day + 14 h Weekend	6 hours/day + 14 h Weekend
Hours per week	53	58	63	68	73	78	83
Costs per week							
Basic rate	95.40	104.40	113.40	122.40	131.40	140.40	149.40
Time & a half overtime	3.60	8.10	12.60	17.10	21.60	21.60	21.60
Double time overtime	18.00	18.00	18.00	18.00	18.00	27.00	36.00
Costs not related to hours	60.99	60.99	60.99	60.99	60.99	60.99	60.99
Total subject to GNI	177.99	191.49	204.99	218.49	231.99	249.99	267.99
Graduated National Insurance	24.38	26.23	27.40	27.40	27.40	27.40	27.40
Costs not subject to GNI	24.70	24.70	24.70	24.70	24.70	24.70	24.70
Total cost per week	227.07	242.42	257.09	270.59	284.09	302.09	320.09
Total cost per hour	4.28	4.18	4.08	3.98	3.89	3.87	3.86

Figure 23.3 Payroll costs per man hour for a seven-day week

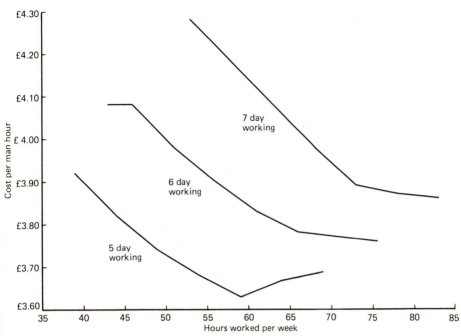

Figure 23.4 Payroll cost per man hour for various working hours

COSTS SUBJECT TO GNI	5 DAY WEEK	6 DAY WEEK	7 DAY WEEK
BONUS	30.00	36.00	42.00
TRAVEL TIME	7.20	8.64	10.08
MAINTENANCE TIME	0.90	1.08	1.26
WET TIME	3.27	3.27	3.27
PUBLIC HOLIDAYS	2.55	2.55	2.55
TOOL ALLOWANCE	0.20	0.20	0.20
SICK PAY*	0.63	0.63	0.63
MISCELLANEOUS	1.00	1.00	1.00
TOTAL SUBJECT TO GNI	45.75	53.37	60.99
COSTS NOT SUBJECT TO GNI			
FARES	9.45	11.34	13.23
SUBSISTENCE	2.47	2.47	2.47
HOLIDAY STAMPS	9.00	9.00	9.00
TOTAL NOT SUBJECT TO GNI	20.92	22.81	24.70

* Sick pay subject to GNI from 6 April 1983

Figure 23.5 Costs per week of payroll items that are not directly related to number of hours worked

Number of working weeks per year

Although the calendar year consists of 52 weeks the number of weeks worked by each man on the site will be considerably less than this. Holidays, both public and annual, currently absorb almost six weeks, sickness may take a further week and two weeks may be lost for inclement weather, a total loss of nine weeks leaving only 43 working weeks on which to set aside funds to pay for this lost time, e.g. sick pay, wet time, etc.

Bonus level

The guaranteed minimum bonus laid down by the Working Rule Agreement is a fixed sum per week regardless of overtime worked. Certainly the expectancy of workmen to bonus is related more to a sum per week than to a rate per hour. How much this sum should be will vary depending on the area of the country, size of contract, trades involved, etc., but for the purpose of this exercise a weekly sum of £30.00 has been included with an additional £6.00 per day for working Saturday and Sunday.

Travel time allowance

The travel time allowance is based on the straight line distance that the site is situated from the workman's home. Assuming an average distance of 30 km, the first 6 km attract no allowance but the remaining 24 km are paid at 6p/km = £1.44/day., i.e. £7.20/week

Maintenance time

Allowing for one man in 10 receiving 1 h/day allowance for refuelling, servicing, maintenance and repairs to plant a daily allowance is necessary of £1.80 ÷ 10 = 18p.

Wet time

Accepting that a total of two weeks in the year may be lost due to inclement weather the cost of these two weeks must be spread over the 43 working weeks of the year, i.e. 2 weeks × 39 h @ £1.80/h = £140.40/year ÷ 43 = £3.27/week.

Public holidays

Eight days of public holidays are recognised in the construction industry and are paid for directly by the contractor at the number of working hours appropriate to that day, i.e. 5 days @ 8 h + 3 days @ 7 h for Christmas Day 1981 and New Year and Good Friday 1982, a total of 61 h @ £1.80 = £109.80/year ÷ 43 working weeks = £2.55/week.

Tool allowance

Allowances for tools vary from 25p per week for bricklayers to 50p for joiners with no payment for labourers. An average has been taken of 20p per week.

Sick pay

The daily allowance for sick pay is currently £4.49 which, assuming one week's sickness per year, amounts to £4.49 × 6 days including Saturday = £26.94/year ÷ 43 working weeks = 63p/week.

Miscellaneous payments

A variety of small payments may be made for such duties as writing out ganger's allocation sheets, driving the minibus to work, pay in lieu of notice, etc. An allowance of £1.00 per week has been made to cover these payments.

Fares

Based on the same distance as the travel time allowance the payment for fares for workmen using public transport would be £1.89/day, i.e. £9.45/week. It is assumed that any cost using company buses for transporting men to work would be roughly equal to this fares allowance.

Subsistence

Allowing for one man in 20 living away from home, a subsistence allowance is necessary of £49.35/week ÷ 20 = £2.47/week. The cost of travel time and fares for periodic weekend leave have been ignored in Figure 23.5 but may warrant inclusion where a high percentage of men are living away from home.

Holiday stamps

Annual holidays and death benefit are funded from a weekly contribution by the employer of £9.00/man providing the man has worked four days in the week. The stamp value is actually £9.10, the remaining 10p coming from the surplus already in the fund.

Redundancy and CITB levy

For the purpose of this exercise the cost of redundancy and CITB levy have not been considered as part of the payroll. Their cost, however, together with other fixed costs of an overhead nature would benefit from an increase in overtime worked.

Graduated National Insurance

On earnings over £27.00/week and up to a maximum of £200.00 an employer must pay GNI contributions of 13.7%. Some of the costs shown in Figure 23.5 are subject to GNI and others are not. These have therefore been totalled and transferred separately into the calculations shown in Figures 23.1 to 23.3.

Basic rate

Assuming an equal number of tradesmen and labourers the average basic rate based on £1.90 for tradesmen and £1.62½ for labourers is equal to £1.76¼. Taking an average plus rate of 3¾p the basic rate for the site including plus rates thus works out to £1.80/h.

Overtime

The national working rules for the building industry currently sets normal working hours of 39 h/week, i.e. 8 h Monday to Thursday and 7 h on Friday. Any hours worked beyond this on a weekday are paid at time and one half. Beyond three hours of overtime (four hours in the Civil Engineering Working Rule Agreement) in any one day are paid at double time. Weekends are paid at time and one half until 4 pm (12.00 noon or after four hours' work in the Civils WRA) on Saturdays and then at double time afterwards. As civil engineering contracts tend to work more overtime than building contracts the example is based on the Civil Engineering Working Rule Agreement.

Plant

The problem of course does not end with only the hourly costs of the workforce. If a contract is plant intensive and much of that plant is hired by the week rather than by the hour then it appears more profitable to work that plant as many hours per day as possible because during overtime hours only the running costs of the plant generate increased expenditure.

Overheads

A similar situation exists with the cost of fixed overheads on the site which would benefit from an increase in hours worked.

Output

Conversely, however, as hours of work are increased so performance of the workforce will drop and this lower level of output must be considered before any decision is made to work consistent overtime.

Cost of absenteeism

It is often believed that because a man is not paid when he is not at work that he therefore does not cost his employer any money. This is not true. As the previous exercise has shown there are a number of irregular payments and costs that must be made during the course of a contract. Funds for this must be set aside during the hours and weeks actually worked.

Calculations in this chapter (as discussed on page 216) assume a working year of 43 weeks, i.e. allowing a nominal one week of sickness per year. This may, however, be further reduced by absenteeism and prolonged sickness, it being not unusual to find construction sites where men are working, on average, for less than 40 weeks out of the 52.

Looking at these costs from an annual point of view it will be found that a man requires an expenditure of several hundred pounds which are not related to actually working. This expenditure is made up from the cost of

- Public holidays
- Days rained off
- CITB levy
- Redundancy

which can be calculated as follows

Fixed annual payroll costs of employing a man
Cost of public holidays

(1) Workmen not on subsistence

	Per week
Guaranteed minimum earnings	
Tradesmen	£86.19
Labourers	73.51½
Average, say 50 : 50	£79.85
Graduated National Insurance on above @ 13.7%	£10.94
Holiday with pay stamps during public holidays	9.00
Total cost for 5 days	£99.79

No. of public holidays/year = 8 days
∴ Cost/year = £99.79 ÷ 5 × 8   £159.66/year

(2) Workmen on subsistence

Cost for 5 days as above	£99.79
Subsistence allowance/week	49.35
Periodic travel time expenses say £30.00 each 6 weeks	5.00
Total cost for 5 days	£154.14

∴ Cost/year £154.14 ÷ 5 × 8   £246.62/year

Cost of wet time

(1) Workmen not on subsistence

Average guaranteed minimum earnings as above	£79.85
Travel time allowance 5 days @ say 30 km	
i.e. 5 days × 24 km @ 6p/km	7.20
	£87.05

Graduated National Insurance on above @ 13.7%	£11.93
Holiday with pay stamps whilst rained off	9.00
Fares allowance 5 days @ say 30 km i.e. 5 days × £1.89	9.45
Total cost for 5 days	£117.43
Assuming total of 10 days rained off/year ∴ Cost/year = £117.43 ÷ 5 × 10	£234.86

(2) Workmen on subsistence (assume digs less than 6 km from site)
Cost/5 days will be same as for public holidays

Total cost for 5 days	£154.14
∴ Cost/year £154.14 ÷ 5 × 10	£308.28

CITB levy

Levy proposed for 1981 (per year)

Tradesmen	£62.00
Scaffolders	£42.00
Skilled workers	£23.00
Labourers	£15.00
Labour-only sub-contractors	2%

Say average per man of £40.00/year

Redundancy

This depends on age and length of service but taking a 35-year-old with five years' service as being average, then four weeks' pay would be due as redundancy (see Department of Employment tables and booklet, HMSO). Thus if pay as defined by 'The Redundancy Payments Scheme' averaged the current maximum of £130 per week during the 12 weeks prior to pay off, then £520 would be due. The Government however currently contribute 41% of this sum leaving a cost of 306.80 to spread over the five years that the man was employed

£306.80 ÷ 5 years = £61.36/year

Summary of fixed annual payroll costs

	Per year	÷ 43 = per week	÷ 39 = per hour
(1) Men not on subsistence			
Public holidays	£159.66	£3.71	9.5p
Wet time	234.86	5.46	14.0p
CITB levy	40.00	0.93	2.4p
Redundancy	61.36	1.43	3.7p
	£495.88	£11.53	29.6p

Amount of overtime worked

(2) Men on subsistence

Public holidays	£246.62	£5.73	14.7p
Wet time	308.28	7.17	18.4p
CITB levy	40.00	0.93	2.4p
Redundancy	61.36	1.43	3.7p
	£656.26	£15.26	39.2p

Workmen not on subsistence						
No. of days sick during week		5 days	4 days	3 days	2 days	1 day
Annual costs (÷ 43 weeks)						
Public holidays	£159.66	3.71	2.97	2.23	1.48	0.74
Wet time	£234.86	5.46	4.37	3.28	2.18	1.09
CITB levy	£ 40.00	0.93	0.74	0.56	0.37	0.19
Redundancy	£ 61.36	1.43	1.14	0.86	0.57	0.29
		11.53	9.22	6.93	4.60	2.31
Weeks costs						
HWP stamp	£ 9.00	9.00	7.20	5.40	3.60	1.80
Sick pay (6 days)	£ 26.94	26.94	17.96	13.47	8.98	4.49
GNI on sick pay (after April 6, 1983)		–	–	–	–	–
Cost for week		47.47	34.38	25.80	17.18	8.60

Workmen on subsistence						
No. of days sick during week		5 days	4 days	3 days	2 days	1 day
Annual costs (÷ 43 weeks)						
Public holidays	£246.62	5.73	4.58	3.44	2.29	1.15
Wet time	£308.28	7.17	5.74	4.30	2.87	1.43
CITB levy	£ 40.00	0.93	0.74	0.56	0.37	0.19
Redundancy	£ 61.36	1.43	1.14	0.86	0.57	0.29
		15.26	12.20	9.16	6.10	3.06
Weeks costs						
HWP stamp	£ 9.00	9.00	7.20	5.40	3.60	1.80
Sick pay (6 days)	£ 26.94	26.94	17.96	13.47	8.98	4.49
GNI on sick pay (after April 6, 1983)		–	–	–	–	–
Subsistence	£ 49.35	49.35	39.48	29.61	19.74	9.87
Periodic travel	£ 5.00	5.00	4.00	3.00	2.00	1.00
Cost for week		105.55	80.84	60.64	40.42	20.22

Figure 23.6 Direct additional cost of sick leave

These annual costs can now be added to the costs incurred on the day that the workman is actually absent as illustrated in Figure 23.6 which shows the cost of employing a man whilst he is absent because of genuine sickness.

Figure 23.7 illustrates the cost of a man absent without acceptable reason and it has been assumed that if he were absent for more than one day that he would not qualify for a holiday with pay stamp that week thus providing some saving to the company. It is also possible that a saving on GNI would occur where earnings fall below the minimum limit of £27.00/week. This

Workmen not on subsistence					
No. of days absent during week	5 days	4 days	3 days	2 days	1 day
Annual costs as for sickness	11.53	9.22	6.93	4.60	2.31
Week's costs					
HWP stamp £9.00	–	(– 1.80)	(– 3.60)	(– 5.40)	1.80
Cost for week	11.53	7.42	3.32	(– 0.79)	4.11

Workmen on subsistence					
Annual costs as for sickness	15.26	12.20	9.16	6.10	3.06
Week's costs					
HWP stamp £9.00	–	(– 1.80)	(– 3.60)	(– 5.40)	1.80
Subsistence (stopped when absent)	–	–	–	–	–
Periodic travel £5.00	5.00	4.00	3.00	2.00	1.00
Cost for week	20.26	14.40	8.56	2.70	5.86

Figure 23.7 Direct additional cost of absenteeism

possibility has not been reflected in the calculations as some men will earn above this amount even when working for only one day. Figure 23.8 looks at the cost of poor punctuality by calculating the cost of each hour a man is late.

It must be stressed that these calculations have only attempted to assess the direct losses in terms of payroll expenditure. The disruption caused by

WORKMEN NOT ON SUBSISTENCE		
ANNUAL COSTS (÷ 43 weeks ÷ 39 h)		PENCE/1 H LATE
Public holidays	£159.66	9.5
Wet time	£234.86	14.0
CITB levy	£ 40.00	2.4
Redundancy	£ 61.36	3.7
		29.6p
WEEK'S COSTS (÷ 39 h)		
HWP stamp	£ 9.00	23.1
Fares allowance	£ 9.45	24.2
Travel time allowance	£ 7.20	18.5
GNI on travel time	+ 13.7%	2.5
COST FOR EVERY 1 H LATE		say 98p

WORKMEN ON SUBSISTENCE		
ANNUAL COSTS (÷ 43 weeks ÷ 39 h)		PENCE/1 H LATE
Public holidays	£246.62	14.7
Wet time	£308.28	18.4
CITB levy	£ 40.00	2.4
Redundancy	£ 61.36	3.7
		39.2p
WEEK'S COSTS (÷ 39 h)		
HWP stamp	£ 9.00	23.1
Subsistence	£ 49.35	126.5
Periodic travel	£ 0.83	2.1
COST FOR EVERY 1 H LATE		£1.91

Figure 23.8 Direct additional cost of poor punctuality

absenteeism and poor punctuality add yet further to such costs making it essential that employers encourage full attendance from all employees and set up disciplinary procedures for consistent offenders.

Hours effectively at work

Chapter 17 introduced the work study techniques of activity sampling. Figure 23.9 illustrates such a study taken at one-minute intervals around start, finish and break times. Because the study is attempting to establish a

CONTRACT _____ STUDY NO __15__ SHEET NO _1_ OF _1_

Description of work __Number of men at workplace__ Date 23 March 82

Time	No. of men	Time	No. of men	Time	No. of men	Time	No. of men	Time	No. of men	Time	No. of men
8·00	Nil	9·30	36	10·15	Nil	12·30	34	1·30	Nil	4·30	35
·01	Nil	·31	36	·16	Nil	·31	34	·31	Nil	·31	35
·02	Nil	·32	36	·17	Nil	·32	34	·32	Nil	·32	35
·03	Nil	·33	36	·18	Nil	·33	36	·33	Nil	·33	35
·04	3	·34	36	·19	Nil	·34	36	·34	Nil	·34	35
·05	3	·35	36	·20	Nil	·35	36	·35	4	·35	35
·06	5	·36	36	·21	Nil	·36	36	·36	4	·36	35
·07	9	·37	35	·22	Nil	·37	35	·37	12	·37	35
·08	9	·38	35	·23	6	·38	35	·38	15	·38	34
·09	15	·39	35	·24	14	·39	35	·39	15	·39	34
·10	18	·40	35	·25	14	·40	35	·40	21	·40	34
·11	24	·41	35	·26	14	·41	33	·41	23	·41	32
·12	31	·42	35	·27	17	·42	33	·42	23	·42	32
·13	35	·43	35	·28	23	·43	33	·43	28	·43	32
·14	35	·44	35	·29	25	·44	33	·44	28	·44	29
·15	35	·45	35	·30	29	·45	33	·45	28	·45	29
·16	35	·46	35	·31	29	·46	28	·46	28	·46	27
·17	35	·47	32	·32	34	·47	28	·47	28	·47	27
·18	35	·48	32	·33	34	·48	28	·48	34	·48	27
·19	36	·49	32	·34	34	·49	25	·49	34	·49	25
·20	36	·50	26	·35	34	·50	25	·50	36	·50	21
·21	36	·51	24	·36	36	·51	18	·51	36	·51	17
·22	36	·52	24	·37	36	·52	18	·52	36	·52	17
·23	36	·53	18	·38	36	·53	13	·53	36	·53	12
·24	36	·54	12	·39	36	·54	9	·54	36	·54	9
·25	36	·55	12	·40	36	·55	9	·55	36	·55	1
·26	36	·56	2	·41	36	·56	3	·56	36	·56	Nil
·27	36	·57	2	·42	36	·57	Nil	·57	35	·57	Nil
·28	36	·58	Nil	·43	36	·58	Nil	·58	35	·58	Nil
·29	35	·59	Nil	·44	36	·59	Nil	·59	35	·59	Nil
·30	35	10·00	Nil	·45	36	1·00	Nil	2·00	35	5·00	Nil

Figure 23.9 Activity sample at one-minute intervals around start to finish times

pattern to the day, it is preferable to taking readings at regular intervals and not at random times as is usually the case in an activity sample. Such a study may well have to be carried out section by section as it may not be possible to reach all areas of the site in the one-minute intervals suggested. This is, however, preferable to taking such samples over a longer period as the difference in the number of men at the workplace in, say, a five minute interval may be the difference between, e.g. 8.05 am nil, 8.10 am 100% present.

Using the example shown in Figure 23.9, the number of men observed during the first 30 minutes totals 722 which when divided by the 30 one-minute readings gives an average of 24 men at the workplace between 8.00 am and 8.30 am. Had all 36 men been there for the full half hour then the man minutes would have totalled $36 \times 30 = 1080$. The number of man minutes lost therefore is $1080 - 722 = 358$ man minutes. Put a different way, the average time of arrival at the workplace was 358 man minutes lost $\div$ 36 men = 10 minutes lost after 8 o'clock, i.e. 8.10 am.

The lost time at other periods of the day can be similarly calculated and the effective working hours in the day determined. Where the effective working hours in the day are found to be lower than is acceptable, site management must exercise stricter control of the men leaving the canteen, etc.

It is important that the responsibility for 'getting the men out of the cabin' is clearly defined, e.g. between foremen and timekeepers. Foremen must set a good example and where transport is necessary it must be available on time and not wait for stragglers. Thought may need to be given to portable canteens and toilets that can easily be moved with the work and not have everyone on the site being transported to and from a central set up. On contracts working long hours consideration should be given to having two half-hour breaks instead of a series of smaller breaks throughout the day.

On large sites clocking on should be arranged locally rather than centrally to avoid time being wasted getting to the workplace. It is clearly cheaper for one timekeeper to travel down the site to the men rather than have all the men reporting to a central timekeeper. A good incentive scheme will help as time lost will mean lower production and therefore lower bonuses.

Finally it must be remembered that the payment of time spent on meal breaks is subject to local agreement and is not laid down in the working rule agreement. In the example shown in Figure 23.9 the breakfast break is paid as working hours but the lunch break is unpaid.

Because this particular site is working until 5.00 pm the last half hour of the day is subject to overtime premium. If the breakfast break were not paid for, then only one quarter of an hour of the day would be classed as overtime. The cost of breakfast break is therefore ¼ hour plus the overtime premium for ¼ hour that it is creating later in the day. In other words, although the breakfast break is being taken during normal working hours its true cost is at overtime rates. Furthermore if another quarter of an hour is lost during the course of the day in arriving late or departing early from the workplace then this too can be costed at time and one half. Looked at a different way, a half hour of overtime is being worked for the benefit of the breakfast break and poor timekeeping at the place of work. Such overtime, because it is non productive, is a total loss and cannot even be considered to help spread the fixed costs discussed previously in this chapter and summarised in Figure 23.5.

Shift working

With the gradual decrease in the working week it is likely that in order to achieve the required progress more sites will consider working two shifts per day. The nature of the work and local restrictions such as night and Sunday working, permitted noise levels, etc., will dictate whether this is possible though much can be done to silence plant in order to make it acceptable especially when the prize is early completion of a contract with possible savings to both client and contractor. It has been shown earlier in this chapter that overtime working can, in certain instances, reduce costs as more hours are worked; however, the falling-off in performance as overtime increases has the opposite effect and is in any case not socially acceptable in these days of high unemployment. The answer may lie in an increase in shift working in the years ahead.

Use of labour-only sub-contractors

It is sometimes beneficial to sub-let work to a labour-only sub-contractor either because of a clear saving in costs or where fluctuation in the workforce is unavoidable. Although this is usually done on a measured rate basis there are frequently instances where a lump sum or rate per hour is agreed.

Whatever the method of payment, however, it is essential that the relationship between the cost of using the sub-contractor and the cost of using direct labour is known. It is all too easy to pick up a telephone and have a labour-only company provide the site's workforce but if the cost of such workmen is higher than can be achieved by using direct labour then clearly consideration must be given to employing labour direct. In making such a comparison, account must also be taken of the annual costs, etc., discussed previously in this chapter as these costs would not be carried by the main contractor if the work were sub-let at all-inclusive rates.

Chapter 24
National increases

Except during relatively short contracts it is likely that some national increase in rates will occur at some time during the course of a contract, i.e. a nationally agreed increase in the basic rates of craftsmen and labourers or in their allowance for fares, travel time, subsistence or sick pay, in the premiums paid for overtime or in the numerous plus rates and extra payments that are laid down by the various working rules*. Changes occur regularly in the cost of the weekly stamps for the industry's annual holiday with pay scheme and death benefit and at unpredictable times the government of the day may impose or cause any number of changes or new costs to be introduced, e.g. Graduated National Insurance and its associated surcharges, redundancy payments, training levies and the now defunct Selective Employment Tax, threshold payments and the industry's own joint board supplement.

Plant costs, though less erratic are also liable to change and include one of the most influential elements of inflation, i.e. fuel costs. Contracts entered into between a client and a contractor may deal with these national increases in a number of different ways.

(1) Fixed price where the builder or contractor agrees to carry out the work for a fixed sum regardless of any price changes whatsoever.
(2) Fixed price where the client is liable only for these increases (or decreases) imposed by the Government.
(3) Fluctuating contract where the client is liable to pay to the contractor any national increases provided the contractor has agreed with his client a list of basic rates that were current at the time the tender was prepared. It is up to the contractor therefore to list any rates that are not nationally laid down if he intends to claim any subsequent increases.
(4) Fluctuating contract where rates priced at the time of tender are increased in line with indices laid down by the Osborne Formula for building works and the Baxter Formula for civil engineering works. (DoE – Property Services Agency for use with National Economic Development Office Price Adjustment Formulae. Indices available from HMSO).

* *National Working Rules*, National Joint Council for the Building Industry. *Working Rule Agreement for the Civil Engineering Construction Conciliation Board for Great Britain.*

Bearing in mind these different methods of allowing for national increases, a decision has to be made on whether cost records for a contract are to be calculated at the actual cost levels current at the moment of construction or at those prevailing at some fixed date either at the start of the contract or when the tender was being prepared.

Looking at the practicalities of this decision it is not always an easy mathematical exercise to calculate what something would have cost if rates had stayed as they were at some previous fixed date especially if bonus is included in that cost. Furthermore the fact that something would have cost

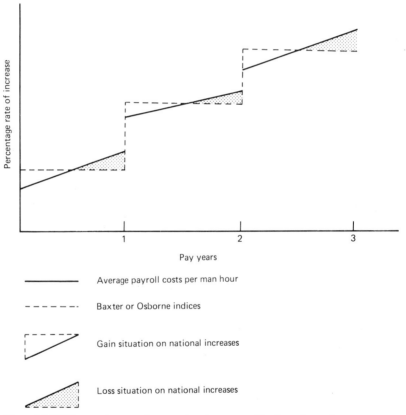

Figure 24.1 Difference between increased costs recovered by formulae and actual expenditure

an assessed £10.00/m³ if rates had stayed as they were 12 or 18 months ago does not have the same ring of truth as the cold fact that it did cost £12.00/m³ when it was carried out. It is recommended therefore that costs are produced at the levels of cost actually incurred at the time of construction.

Where costs are being compared with standards or values that are based on the rates in the Bill of Quantities those rates will be as agreed at the time of tender. An adjustment must therefore be made to include the monies recovered under whatever terms apply for recovery of national increases and/or to include any allowances for such increases made by the contractor within his tender, especially where the contract is let on a fixed price basis.

Usually this is only necessary once a year when rates of pay are reviewed and a straight percentage addition can be added to all unit values and budgets. At that point in time a line can be drawn across the cumulative cost figures if so desired in order to separate one pay year's figures from the next. Alternatively such increases can be considered as an overhead item in its own right, actual cost of the increases being compared with the amount of monies recovered or allowed in the tender.

In the case of contracts relying on formulae for recovery of national increases this comparison is important as losses may well be occurring on a contract because the site is paying increases that are not reflected in the formula e.g. more than average overtime or plus rates, higher than anticipated bonus payments, etc. Any comparison of average actual payroll costs with the indices will reveal a difference in the respective rates of change as the example in Figure 24.1.

The indices jump suddenly at the anniversary of a national increase whereas the actual jump in payroll costs is usually smaller but continues to drift upwards until the next pay round. Of course this illustration is very much simplified as many other factors affect the indices such as increases in GNI and affect the wage costs such as the stage of construction that a particular contract has reached or other contracts starting in the same area. However, if the pattern shown in Figure 24.1 is found in a company's wage records, care must be taken and extra allowances made when pricing contracts at different times of the year, i.e. a contract with base data just prior to a national increase will have the benefit of the jump in index without the corresponding costs, whereas a contract with base data after the national increase may suffer a drift of wages upward without any compensating jump in the index. Over and above this swinging of the index to either side of actual costs may be a more gradual drift over a period of years which, although not so dramatic as the swinging action, can nevertheless eat substantially into the profit margin allowed for a contract.

Chapter 25

Incentive schemes

The recognised system for payment of labour under the Working Rules* of both the building industry and the civil engineering industry is on a basis of the number of hours a man presents himself for work. The working rules however allow for additional payments to be made in the form of bonus, the national working rules for the building industry even lays down a set of general principles concerning incentive schemes and productivity agreements the main points being as follows

(1) Targets should, where possible, be issued before work commences.
(2) Targets should be based on work study principles published by the British Standards Institution.
(3) The number of men treated as a unit for bonus purposes is to be as small as possible, certainly not calculated on a trade or site basis.
(4) The incentive scheme should be expressed in simple and precise terms so that the men can understand the scheme and avoid misunderstandings.
(5) Pre-measured tasks should be as short as possible.
(6) Targets must include for both tradesmen and labourers.
(7) Overtime and travel time payments are not to be charged against the targets.
(8) Bonus earnings should be notified to the workmen as soon as possible.
(9) Defective work should be paid for by the gang who carried it out.

What then are the advantages and disadvantages of operating such a scheme which by definition will result in costs per man hour being higher than those laid down by the industries' wage negotiating bodies.

Advantages of a measured incentive scheme

(1) Outputs achieved will generally be better than if payment were based only on hours worked. This can result from an increase in effort

---

* National Working Rules for the Building Industry.
Working Rule Agreement for the Civil Engineering Construction.
Conciliation Board for Great Britain.
BATJIC agreement for smaller builders.

and/or an improvement in the way each man organises himself, e.g. in making sure that tools and materials are collected from the stores without having to make a special journey or in completing a job before having a meal break or leaving at night. In getting on with the next job when one job is finished and in alerting management more rapidly to the fact that he is waiting for work or is held up for some reason. The outcome of this is usually a reduction in unit costs and an improvement in progress with a resultant gain in overheads.

(2) The additional payments will generally attract the more industrious type of labour who is prepared to work harder for higher reward. Furthermore in times of scarcity of skilled labour only those companies making such additional payments are likely to retain the limited labour that is available thus helping to reduce labour turnover.

(3) The data used to operate the bonus scheme provide a variety of statistics that assist site management to monitor the running of their contract. Such output, performance and cost statistics would probably not otherwise have been collected.

Disadvantages of a measured incentive scheme

(1) A heavier burden falls on management to see that the rush by some gangs to complete work as quickly as possible and thereby earn higher bonuses does not lead to shoddy workmanship and falling standards of safety.

(2) Additional costs are involved in someone setting targets, measuring work and calculating bonuses to a reliable degree of accuracy.

(3) Disagreements, disputes and possibly strikes may result from the operation of the scheme, over the targets being used and over what work is included in the targets.

Type of scheme

A number of different systems of incentive schemes are in use, the type of scheme tending to vary in different areas of the country and between different companies, local authorities, etc. Schemes can be grouped under the following six headings which are summarised in Figure 25.1 to show the main advantages and disadvantages of each.

(1) Profit sharing
(2) Piecework
(3) Time saved
(4) Performance related
(5) Assessed bonus
(6) Job and finish

Profit sharing schemes

This type of scheme, where an element of the audited profits of a company are paid out as a bonus, is generally too far removed from the efforts of the

Incentive schemes

Type of schemes	Advantages	Disadvantages
Profit sharing	Easy to operate without additional staff	Too remote from 'shop floor' to reflect efforts of workmen
Piecework	Easily understood Standard times are easily cashed out for use as targets	Difficult to gear
Time saved	Easily geared by changing relationship between standards and targets or by changing saving rate Different saving rates can be used for, e.g. any men without full attendance in week	Difficult to convert targets from standards at differing saving rates to give chosen bonus levels
Performance related	Can be set to quite complex gearing with different tables for, e.g. degree of material wastage Standards are used direct	Not easy to understand
Assessed bonuses	Does not require additional staff to operate	Depends on subjective assessments. No guarantee that the men are giving value for money
Job and finish	Good motivation Easy to operate where a number of men are involved in a single task, e.g. large pour of concrete	Workmen's transport home has to be arranged for the job and finish gang. Not possible where a number of different tasks and gangs are involved

Figure 25.1 Different types of incentive schemes

workman on site to be considered for anything other than an incentive scheme for supervisory labour and staff. Profit and loss of a particular site and of the company as a whole may be due to commercial reasons over which the workmen have no control.

Piecework schemes

The payment of piecework rates, e.g. laying facing bricks £60/thousand is the easiest system to understand. The workman knows that for every 1000 bricks he lays he gets £60 in basic pay from which his basic hourly earnings will be deducted to calculate his bonus. Take as an example a gang of three bricklayers and one labourer who have laid 7000 facing bricks between them in a week of 39 h.

Piecework earnings = 7000 bricks @ £60/1000	£420.00
Basic hourly earnings	
3 tradesmen × 39 h @ £1.90* = £222.30	
1 labourer × 39 h @ £1.62½ = £63.37½	£285.67½
∴ Bonus earnings to be shared between the gang	£134.32½

Targets in a piecework scheme are often set at the 'going rate' for a particular area especially within the bricklayer trade.

It is important, however, if disputes are to be avoided, to take more care in setting targets and to resort to the use of work study techniques. Having established standard times for items of work it is then relatively easy to convert these times into piecework targets. This can be done by adding a policy allowance to the standard time as follows

Standard time for hanging internal hollow core door say 0.50 h/No.

Add say 33⅓% *policy allowance* to allow bonus of ⅓
   of basic rate to be earned at 100 performance   0.17 h/No.

Allowed time for hanging internal hollow core door   0.67 h/No.

Piecework target = 0.67 h/No. @ basic rate £1.90
                  = £1.27/door

Alternatively a bonus level can be set for achieving 100 performance and used together with the basic rate for cashing out standard times.

Tradesmen's basic rate	£1.90/h
Bonus to be paid for 100 performance, say	£0.70/h
Rate for cashing out standards	£2.60/h

Standard time for hanging internal hollow core door, say 0.50 h/No.
Piecework target = 0.50 h/No. @ cashing out rate £2.60
                  = £1.30/door

The piecework rate for brickwork can similarly be built up by either of these methods.

Assuming a 3 : 1 bricklayer gang the average basic rate would be

3 bricklayers @ £1.90	£5.70
1 labourer @ £1.62½	£1.62½
—	
4	£7.32½
Average basic rate	£1.83

* At 1981/82 rates

Standard time for laying facing bricks, say 20.00 h/1000

Add say 66⅔% *policy allowance* to allow
bonus of ⅔% of basic rate to be earned at
100 performance   13.33 h/1000

Allowed time for laying facing bricks   33.33 h/1000

Piecework target = £33.33 h/1000 @ average basic rate £1.83
                 = £60.99/1000

Equally the target could be calculated by the second method

Average basic rate	£1.83/h
Bonus to be paid for 100 performance, say	£1.17/h
Rate for cashing out standards	£3.00/h

Standard time for laying facing bricks, say 20 h/1000
Piecework target = 20 h/1000 @ cashing out rate £3.00/h
                 = £60.00/1000

The advantage of the second method is that it avoids confusion between standard times and allowed times and also helps management to see the level of bonus that they are planning to pay. This bonus level can of course be expressed as a percentage of the basic rate should a direct link with the basic be required.

An extension of this system is used in the payment of labour-only sub-contractors carrying out work on a measured basis. The comparable rate paid to such a sub-contractor would be increased to say £80/1000 to include for such wage sheet costs as Graduated National Insurance, holidays stamps, travel time or fares. The amount of the payment to the sub-contractor is then even easier, the only deductions necessary being any CITB levies which have to be paid by the main contractor, general insurances and retention that may be withheld.

Time-saved schemes

Probably the most common scheme in the construction industry is the time-saved system. Output targets are compared with actual time taken and any hours saved against that target are paid at a predetermined rate per hour. For example, taking the same bricklayer gang as before but on a time-saved target of 30 bricks/h

Hours earned = 7000 bricks ÷ 30 bricks/h	233 h
Hours taken = 4 men @ 39 h/week	156 h
Time saved	77 h

If each hour saved is paid as bonus at the average basic rate of the gang, i.e. £1.83, then bonus earned will be 77 h saved @ £1.83 = £140.91 bonus.

Often with time-saved schemes, however, the hours saved are not paid at the full basic rate, either being calculated at a fixed rate, e.g. hours saved paid at 75p/h or at a percentage of the basic rate, e.g. hours saved paid at 50% of basic rate. The targets are correspondingly adjusted to allow bonus levels to be maintained, e.g. at a time-saved target of 20 bricks/h and hours saved paid at 75p/h the above calculations would now be

Hours earned = 7000 bricks ÷ 20 bricks/h   350 h
Hours taken = 4 men @ 39 h/week   156 h

Time saved   194 h
194 h saved @ 75p = £145.50 bonus

As with piecework schemes it is prudent to develop targets from standard times. This can be done by adding a policy allowance as before but with an adjustment if the hours saved are to be paid at a rate other than the basic rate thus the allowed time for hanging internal hollow core doors of 0.67 h/No. previously calculated can be used, without adjustment, as a time-saved scheme target provided the hours saved are to be paid at the full tradesmen's basic rate of £1.90/h.

To adjust this target for paying only 75p/h saved, however, requires the policy allowance to be adjusted by £1.90/75p,

i.e. policy allowance paying £1.90/h saved was 33.33%
Adjustment for paying 75p/h saved

$$33.33\% \times \frac{£1.90}{£0.75}$$

∴ policy allowance paying 75p/h saved = 84.44%

Alternatively this policy allowance can be calculated as

$$\frac{\text{Bonus level required (i.e. ⅓ of basic)} \times 100}{\text{Saving rate} \qquad 1}$$

$$= \frac{63.33p}{75p} \times \frac{100}{1}$$

$$= 84.44\%$$

The target can now be calculated thus

Standard time for handling internal hollow core door, say	0.50 h/No.
Add 84.44% policy allowance as calculated above to allow bonus of ⅓ of basic rate to be earned at 100 performance	0.42 h/No.
Allowed time for hanging internal hollow core door	0.92 h/No.

It must be noted that in all cases the policy allowance is an addition to the standard time and not a factor to be applied to the standard time. This confusion can arise where output standards are expressed in units per hour instead of hours per unit. For instance to add a policy allowance of say 75% to a standard output for brickwork of say 50 bricks/h, the output must first be inverted as follows

Standard time for laying bricks/1000   1000/50   20 h/1000
Add 75% policy allowance   15

Allowed time for laying bricks/1000   35 h/1000
∴ Target output = 1000/35 h = $28\frac{1}{2}$ bricks/h

It can be seen from the calculations above that both the bonus level required and the saving rate are going to have a dramatic effect on the policy allowance and that the difference between the *standard* time for a job and the *allowed* target time can be considerable. This is often not recognised by the workforce when comparing targets paid on one contract with targets paid on another. It is imperative, therefore, that anyone responsible for incentive schemes fully understands the above formula and can explain its effect to the workforce when challenged with targets from other sources. This applies equally whether or not this type of scheme is currently being used, as targets being quoted will often have derived from a time-saved scheme.

Performance related scheme

A scheme favoured by local authorities and some construction companies is one which uses the standard times direct and calculates actual performances achieved. Bonus is then paid according to a predetermined scale which may increase for each point of performance or for bands of, say five points. Different increments may be used beyond, say 120 performance and different scales can be used for, say workmen with a good record for absenteeism or wastage of material. The scheme is not, however, easy for the operatives to understand and calculate bonuses for themselves.

Taking the previously used example laying facing bricks for which a standard time of 20 h/1000 was assumed, the laying of 7000 bricks would amount to

$$7000 \times \frac{20}{1000} = 140 \text{ standard hours}$$

of work being completed. As the gang took

4 men × 39 h = 156 h to lay the bricks

their performance can be calculated as

$$\frac{\text{Standard hours}}{\text{Actual hours}} \times \frac{100}{1}$$

$$\frac{140}{156} \times \frac{100}{1} = 90 \text{ performance}$$

In the scale shown in Figure 25.2 this would represent a bonus level of 90p/man hour, i.e. a total for the gang of 156 h @ 90p = £140.40. The use

| | Bricklayers and bricklayers' labourers | |
Performance achieved	100% attendance	Any absenteeism during appropriate week
Below 80	GMB	GMB
80–84	60p	40p
85–89	75p	50p
90–94	90p	60p
95–99	£1.05	70p
100–104	£1.20	80p
105–109	£1.35	90p
110–120	£1.50	£1.00
Over 120	1p extra for each point above 120	1p extra for each point above 120

Figure 25.2 Bonus level per hour worked for performance achieved

of wide bands of payment levels assists in the running of the scheme not only because it helps to smooth out fluctuations in earnings from week to week but also because it tends to cut down on the number of minor queries such as, for instance, claims for being delayed for half an hour waiting for a grid line to be set out. The chances are that unless the gang's performance is on the borderline between one band and the next that such an allowance would not make a difference to the gang's bonus earnings and will therefore not be pursued.

Assessed bonuses

There are basically two alternatives for paying bonuses without resorting to measurement of the work and the hassle of setting targets.

Fixed bonus or standing bonus may be paid where a set amount is paid per hour, per shift or per week. This tends to be used more where the progress of the work is outside the control of the workmen such as providing an attendance on others, e.g. sweeping roads, flagman, etc.

Spot bonus or management bonus may be paid at the discretion of site management depending on the foreman's appraisal of each man's effort for the appropriate week. This can only be a subjective assessment and puts a burden of responsibility on the foreman or manager that frequently causes him to resort to paying fixed or standing bonuses albeit perhaps with a few pence variation from week to week as a token gesture towards making such an appraisal. Management bonus is sometimes shadowed or checked by a measured scheme to assist the manager in his assessments.

Job and finish

Where a set amount of work has to be completed during a day then a set number of hours' pay can be agreed and the workmen allowed to go home whenever that work is finished. This can provide a strong incentive on such

238 *Incentive schemes*

work as mass pours of concrete, pipeline work or any tasks where the amount of work that a gang can do is governed by availability, access or programme over which the men have no control, and when they have completed their work for the day there is nothing to be gained from finding them alternative work even if they manage to finish within normal working hours. This system is usually only used for occasional urgent tasks that require an additional push in order to achieve a tight programme.

Depending on the statistics arising from the company's payroll system, appropriate care must be taken in entering such unworked hours on to the payroll as an amount of bonus and not as working hours.

Gearing

An incentive scheme that produces basic earnings, i.e. basic rate and bonus, that are directly proportional to output performances is known as a *straight proportional* scheme. A piecework scheme is usually operated as a

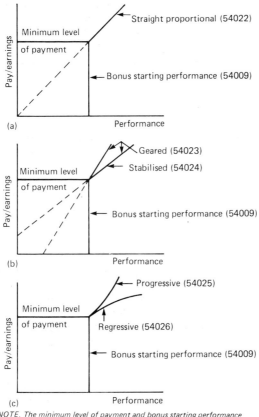

Figure 25.3 Graphical representation of main forms of payment by results

straight proportional scheme, e.g. lay bricks £75/1000. If 10 000 bricks are laid, earnings are 10 000 × £75/1000 = £750. If no bricks are laid earnings are nil (subject to the guaranteed minimum laid down by the appropriate Working Rule Agreement).

Wherever earnings are not directly proportional to output performance as in a time-saved scheme paying a saving rate that is different from the basic rate, then that scheme is said to be geared. Figure 7 of BS 3138:1979, illustrated in Figure 25.3, shows the various types of gearing that are possible, the most common being a *stabilised* scheme where only part of any savings made are paid to the operative as with the previous example of a time-saved scheme paying only 75p/h saved. Such a scheme having the advantage that targets can be made easier and therefore the scheme appears more attractive. On the other hand a rate higher than the current basic rate may be used for payment of hours saved in an attempt to make savings on overheads by attracting the most efficient labour.

Where a minimum is placed on the level of bonus earnings this is referred to as a *fall back* or *guaranteed minimum*. The Joint Board agreement for the building and civil engineering industry currently lays down such a guaranteed minimum though this is not featured in the BATJIC agreement. A maximum level of bonus earnings is referred to as a ceiling and is not generally to be recommended as a ceiling on effort is contrary to the aims of an incentive scheme. Ceilings usually arise where targets have not been set by proper work study techniques and have therefore been found to be slack allowing earnings to be made in excess of those intended.

This slackness of the targets may be the result of insufficient study of the tasks or may be the result of negotiation. It should be borne in mind, however, when entering into negotiations on targets, that the standard times should be either right or wrong but never negotiable. Policy allowances, bonus levels, rate per hour saved, etc., are the negotiable elements of an incentive scheme.

Monitoring incentive schemes

Incentive schemes are like a garden, they continually require attention to keep in trim. This attention must be given at all levels of the company from top management to chargehands level by monitoring the various facts and figures that arise from the scheme. Furthermore, where an incentive scheme is in operation, queries will inevitably arise as to why some gangs or men are earning more than others. In such situations it is useful to be able to refer to statistics to support the superiority or otherwise of the individuals in question. There are three factors that can assist here

- (1) record of bonus earnings. A weekly record of the bonus earnings of a particular gang will show them up as generally being high or low earners and will rapidly expose an abnormal situation if they stray from their normal span of earnings in any week
- (2) output achieved. Although influenced by the nature of the work carried out there should always be some relationship between bonus

earnings and output achieved in any particular week. Where all the work carried out by any gang is substantially similar then the jobs carried out by that gang can be measured using a common unit of measurement, e.g. concrete gangs' work in cubic metres, shuttering gangs' in square metres and the output over the week in hours per unit recorded. Hours that have not been measured can either be included in order to calculate an overall output or can be excluded in order not to influence the output data. Whichever of these options is chosen, however, must then be used consistently

(3) performance achieved. Provided incentive scheme targets have been based on work study standards it is possible to convert earnings back to standard hours and thus monitor performances achieved against the norm of 100 laid down in BS 3138:1979. This has the advantage of making comparison possible between gangs carrying out completely different jobs of work and indeed different trades. The accuracy of these performances will, however, only be as good as the standard times on which the targets are based and will rely on strict application of those targets without unwarranted allowances being made.

The conversion of earnings back to standard hours is achieved by reversing the process carried out when calculating the targets. In a piecework scheme this means dividing the total piecework earnings for the gang by the rate originally used to convert the standard outputs into piecework targets.

$$\text{Standard hours} = \frac{\text{Measured piecework earnings}}{\text{Rate used for cashing out standards}}$$

In the example illustrated on page 233 the measured piecework earnings were £420.00 and if the target were based on the output standard shown on page 234 of 20 h/1000 cashed out at £3.00/h, then the total standard hours produced for the gang would be as follows

$$\text{Standard hours} = \frac{£420.00}{£3.00}$$

$$= 140 \text{ standard hours}$$

Similarly with a time-saved scheme the process can be reversed by comparing the target time with the standard time. Taking the second example on page 235 the relationship between the target of 20 bricks/h and the standard of, say 50 bricks/h is 20:50, therefore the hours earned must be multiplied by 20/50 to convert them back to standards, i.e. a factor of 40%.

Standard hours of work done can therefore be calculated as

$$350 \text{ h earned} \times 40\%$$

$$= 140 \text{ standard hours}$$

Performance can now be calculated using the formula

$$\text{Performance} = \frac{\text{Standard hours of work produced}}{\text{Actual hours taken to do that work}} \times \frac{100}{1}$$

$$\therefore \text{Performance} = \frac{140 \text{ standard hours}}{4 \text{ men} \times 39 \text{ h}} \times \frac{100}{1}$$

$$= \frac{140 \text{ standard hours}}{156 \text{ actual hours}} \times \frac{100}{1}$$

$$= \underline{90 \text{ performance}}$$

Any hours spend on work not measured in the incentive scheme are not to be included as 'actual hours taken to do that work' and therefore will not form any part of the calculation of performance.

Further reading

The Principles of Incentives for The Construction Industry, (1981), The Advisory Service for the Building Industry, 18 Mansfield Street, London W1M 9FG.

Exercises

(1) Standard time for hanging external doors to houses, say 0.75 h/door. Use current basic rate of joiner (no labourer involved). Bonus for 100 performance 33⅓% of basic rate, or assume policy allowance on standard time of 33⅓%. Assume straight proportional scheme, i.e. no gearing. What is the target for hanging external doors in each of the following types of scheme?

    (1) Piecework £ _____ /door
    (2) Time saved _____ doors/h or _____ h/door.
    (3) Performance related _____ doors/h or _____ h/door.

(2) Your contract requires a force of 10 joiners but so far you have only been able to attract half this number who are struggling to achieve standard outputs and complaining that 'the bonus targets are no good'. You have set the level of bonus in a piecework scheme at 45p for 100 performance which taking a sample item of formwork to 500 mm × 500 mm columns at a standard time of 0.90 h/m² fix and strip gives a piecework target of (0.90 × £1.90 basic + 45p bonus), i.e. £2.11½/m² fix and strip.

This happens to be within 2p of what is included in your tender. You have tried some local advertising with little response. You know, however, of a gang of four good joiners available immediately but they are asking £3.00/m² for the same item. You have used this gang before and know that they will have no trouble in reaching high performances by high outputs. They are likely to average 120 performance which would give them 92p/h bonus on your piecework target of £2.11½/m² and £2.10/h bonus on their price of £3.00.

There are also five joiners coming available from another site but they have been making quite high earnings on a not very well controlled incentive scheme and will not now start for less than £1.00/h spot bonus. You do not know this gang but are told by the present site management that they are 'quite good and can turn their hands to anything'. What action are you going to take in order to keep to programme and keep your other trades fully employed?

Chapter 26
Job cards

Where the majority of tasks within a contract can be identified, pre-measured and targeted prior to that section of work commencing, a job card system may be set up as illustrated in Figure 26.1. Targets can be expressed in hours or cash, the tasks listed on the card totalling to make a job that will take the gang between half a day and half a week to complete. Job cards that last only a few hours tend to involve more paper than productive work and can therefore lead to the breakdown of the system. On the other hand, too much work on the card becomes just a list of tasks which has to be continuously revised as some of the tasks are completed. Inevitably, however, some jobs will carry over from one week into the next and a few may even drag on for a number of weeks especially if they are jobs held in reserve for wet weather or other periods of delay.

As a solution to this a simple note marked on the job card of e.g. 75% done may be adequate or a re-issue of the card showing the amount of work remaining. However, confusion over responsibility, over the meaning of a cryptic note or even whether a figure scrawled across the card is the amount left to do or is the amount already completed may well arise where a large amount of work is being carried forward. It is then preferable to set up a more formal record as suggested in Figure 26.2 which can be printed on the back of the job card. The breakdown of measurements on the card into lifts, pours, dwellings, etc., as well as the breakdown item by item assists in assessing such part payments rather than taking an arbitrary percentage complete.

A company's recognised chain of command will usually dictate what is the best method of distributing these cards; however, this must not be left to chance and a procedure such as that illustrated in Figure 26.3 should be drawn up stating who is responsible for each stage of the system. The primary purpose of a job card system is to generate increased motivation set up by an incentive scheme

(1) by making sure that each gang knows what is the target for the work currently being carried out and making it easier for him to calculate the effect on his wages of his improved output

(2) by informing the gang of their next job of work so that they may see their way ahead and flow rapidly from one job to the next
(3) by giving the gang a reserve job that they can fall back on should there be some hold up on their scheduled job.

In addition to this enhancement of the incentive scheme a job card system has a number of other advantages.

				JOB CARD NO. 180	
CONTRACT AB Factory		LOCATION Pump room	GANG Smith		
TRADE Joiner	PLANT Nil		DRG NO. 4	REV B	
OUTLINE OF JOB Fix shutters to first floor slab					

CODE	TASKS IN JOB	MEASURE	UNIT	Target rate	Total target
	Shutter suspended floor including all propping (no make)	250	m^2	0.50	125
	Shutter to edge of slab, beam sides and soffite and nib (make measured separately) including all propping.	60	m^2	1.00	60
	Perforated stop ends 150 mm deep to divide slab into four bays	6	m^2	1.50	9
	Fix polystyrene boxes in slab to form holes for services	40	No.	0.10	4
DATE JOB COMPLETED 12 Aug 82			TOTAL TARGET FOR JOB		198 h
FOREMANS SIGNATURE FOR SATISFACTORY QUALITY & COMPLETION OF ALL ABOVE TASKS J. Smith					

Figure 26.1 Job card

On many contracts, hours are lost in snagging, i.e. making good before handover of a dwelling or section of the work. This snagging can be the result of accidental or malicious damage, attendance on plumbing or electrical sub-contractors, natural shrinkage or drying out or it may be work that was never completed in the first place. Job cards can assist in verifying that work has been completed by requiring a foreman's signature before payment is made. The card also records who was responsible for carrying out the work and when it was done so that a gang can be recalled to complete or replace work that is subsequently found to be unacceptable.

TASKS IN JOB	TOTAL MEASURE	W/E 1/8/82 MEASURE THIS WEEK	W/E 1/8/82 MEASURE REMAINING	W/E 8/8/82 MEASURE THIS WEEK	W/E 8/8/82 MEASURE REMAINING	W/E 15/8/82 MEASURE THIS WEEK	W/E 15/8/82 MEASURE REMAINING	W/E MEASURE THIS WEEK	W/E MEASURE REMAINING
Floor	250 m²	75	175	175	—	—	—		
Edge	60 m²	—	60	45	15	15	—		
Stop ends	6 m²	—	6	—	6	6	—		
Boxes	40 No	—	40	—	40	40	—		
TARGET	198	38	160	132	28	28	—		

Figure 26.2 Record of measure if job is not completed within one week ending

(1) Site agent to define methods of working and position of construction joints approximately three weeks before job is to commence

(2) Quantity surveyor to pre-measure work including tasks not required by the Standard Method of Measurement, e.g. kickers, stop ends, etc. approximately two weeks before job is to commence

(3) Production surveyor/work study engineer to set targets and calculate total target for job approximately one week before job is to commence

(4) Planning meeting each Friday morning to decide what work will be tackled during next 10 days, allocate jobs against gangs and decide priorities within these jobs

(5) Foreman will retain cards for his gangs for the week and will issue to each gang leader, at starting time each day, sufficient job cards for two days' work, placed in order of priority

(6) Gang leader is to return *all* job cards whether work is completed or not to the foreman at finishing time each day

(7) Foreman is to inspect work done and sign job card if work is completed to his satisfaction. Job card is then to be passed to bonus office for calculation of bonuses

(8) Gang leaders are to fill in daily time allocation sheets as usual but need only quote the appropriate job card number for all work that is included on the card.

Figure 26.3 Sample procedure for operating job card system

The card thus provides a detailed record of progress on the contract not only for bonus payment purposes but also for the general progressing of the contract. As an aid to short-term planning (see Chapter 27), the cards can be used as the basis of discussion at a planning meeting where frequent decisions have to be made quickly on alternative programmes of action and no time is available to wait or hold up the meeting whilst someone calculates how much work is involved in two alternative runs of drain or two alternative shutters. The job cards display this information immediately.

In the control of costs on site it is such detailed thinking through of the work that can contribute so much towards the effective use of resources.

Exercise

Trade cost data, as discussed in Chapter 5, are showing that on recent housing contracts an average of 50 h/dwelling has been spent on snagging prior to handover of the buildings. This amounts to approximately 5% of all hours spent on these contracts.

What procedure would you adopt in setting up your next housing contract in order to prevent this wastage of manpower?

Chapter 27

Weekly planning

Unless a contract is to be allowed to drift from crisis to crisis achieving completion solely because of the natural chain of events, then some sort of planning and regular appraisal of the progress of the contract is necessary. This point appears to be generally accepted in the construction industry and most contracts lasting more than a few weeks have some form of programme which is regularly marked up to monitor progress.

There is, however, a different side of planning which is the detailed weekly or short-term planning to finalise arrangements before commencing a particular job of work. It is this short-term planning that is too easily left to chance in the construction industry and yet can do so much to oil the wheels of production and so improve both the efficiency of the workforce and the productivity of the site as a whole.

Because of the differing responsibility of the various members of the site management team no one person is usually in a position to know that all of these final arrangements have been made. It is therefore preferable to call all parties together for a weekly planning meeting. This should be held towards the end of the week to discuss the following week's work plus a few extra days' work as a reserve. Each proposed run of drain, each pour of concrete, each lift of brickwork, etc., should be subjected to any of the following questions that are deemed appropriate to the work, bearing in mind always that the smaller the gang of men involved in the task, the higher the motivation. A gang of one man being the most effective. Only by such close scrutiny can the work be thought through before actual construction starts.

Programme

(1) Is the job critical to the contract programme?
(2) Must it be done at any particular time of day, e.g. a concrete slab which has to be power floated after the concrete has started to set.
(3) Is the client likely to make last minute alterations of restrictions, e.g. outstanding drawing queries or adjacent concrete pour not up to strength yet.

CONTRACT						DETAILED PROGRAMME FOR W/E 1 AUGUST 1982		
						FOLLOWING WEEK		
NAMES	MONDAY	TUESDAY	WEDNESDAY	THURSDAY	FRIDAY	MONDAY	TUESDAY	NOTES
Smith 4 Joiners	Fix first floor slab A1 — C6		Strip FF slab A6 — C10		Fix FF slab	C6 — F10	Strip FF slab A1 — C6	
Jones 4 Joiners	Strip duct walls to F line	Fix duct walls lines F — J	Fix pump house bases	Fix duct walls lines F — J	Fix manhole soffites F7 — F13	Fix manhole soffites F7 — F13	Strip pump house bases	Alternative if cast iron puddle pipe for duct wall not recieved on Monday as promised
Leroy 2 Joiners	Strip GF columns L7 — P9	Fix GF columns L9 — P9	Fix GF columns L9 — P9	Fix FF columns A1 — C4	Fix FF columns A1 — C4	Strip GF columns L9 — P9		
Singh 2 Joiners	Fix kickers retaining wall	Fix channel to sluice	Strip kickers wall	Fix first lift retaining wall	Sluice	Strip first lift retaining wall	Fix second lift retaining wall	
NON PRIORITY JOBS SHOWN THUS	—	—	—					

Figure 27.1 Weekly programme of labour resources

Workplace

(1) Is the area for work available and clear?
(2) Is access into the area available and clear?
(3) Is anyone else working or programmed to work in this area or along the access route?
(4) Is any overhead or weather protection required and prepared?
(5) Is any scaffolding required and erected?
(6) Are all necessary lines and levels set out?
(7) Is any special supervision required during construction, e.g. safety, lines and levels, checking, testing, etc?
(8) Does the area need to checked and passed by the clients representative?
(9) Are any special permits to work required in that area, e.g. adjacent power lines?
(10) How soon can work start in this area?

Material

(1) What materials, fixings, etc., are required?
(2) Where are each of the materials at present?
(3) What action is required to get these materials, etc., to the workplace?
(4) If any of the materials are damaged or become damaged what action is to be taken?
(5) When will all materials, etc., be available at the workplace.

Plant

(1) What is the optimum plant for the job?
(2) Is that plant on site and available?
(3) Who else may require that plant?
(4) What alternative plant is available?
(5) What plant needs to be hired?
(6) Is the plant serviced and in good working order?
(7) Is adequate fuel available?
(8) Where will plant be prior to this job?
(9) When can plant be made available?

Labour

(1) What method is best for carrying out the job?
(2) What is the optimum size of gang to carry out this job?
(3) Are labourers needed in attendance?
(4) Which gang is best for this work?
(5) Will attendance by other trades be necessary?
(6) When can selected gang be programmed to carry out this job?

Having ironed out the problems on any of these points and settled on the optimum plant and labour gang then the next item of work can be considered. In practice such a list of questions need only act as a prompter, many of the answers being obvious to all present at the meeting.

By scheduling the jobs allocated to each gang as suggested in Figure 27.1, a picture can be formed of the following week's programme. Gaps in the work load can be filled with additional jobs and if necessary additional labour and plant brought in or released to obtain the optimum use of the resources available. It can be seen that the use of job cards on a contract greatly assists in the judging of how long a particular item of work will take and a programme of work is produced that is much more likely to work out in practice. A programme in this detail is unlikely to be too accurate when simply based on subjective data, e.g. the feeling that this job should take 'a couple of days' and that one 'the best part of a week'.

Account must be taken of the relationship between targets used as a basis of the weekly programme and the times that are likely to be taken to carry out the job. Targets may be in cash in which case they have to be converted into man hours or targets may be designed for a time-saved system of bonusing or may include a policy allowance for bonus purposes. The gang selected for the work may consist of high flyers who consistently produce above-average outputs or the gang may contain a number of apprentices or slower than average workers. In all of these cases the target times need adjustment in order that a realistic period is considered for planning purposes.

Of course work will not be constructed exactly to the programme anticipated. The unexpected will invariably happen, the weather will play havoc, materials will fail to arrive, plant will break down and key men will report sick, but this surely is a reason for eliminating as many of the anticipatable hold ups as possible so that less is left to chance and alternative courses of action have already been aired.

Site managers will frequently employ a few spare men to cover for absentees and other unknowns. However, unless a costly item of plant is involved this is expensive insurance. The opposite is far more cost effective, i.e. to have *work looking for men* and not *men looking for work*.

Chapter 28

Pre-costing

By setting standards against the contract's weekly programme and comparing the total of these standards with the probable wage bill for the same week it is possible to calculate the profitability of the following weeks' programme even if that programme is not a written one and exists only as the thoughts of the foreman or manager. Thus site management can visualise how essential it is to complete a programme. If the planned programme indicates a loss, it is not too late to study that programme together with those responsible for carrying out the work, so that foremen and gangers can appreciate how near the bone the programme is – how essential it is, for instance, that they achieve that extra pour of concrete, strip that shutter a day earlier or get that JCB off site by Wednesday at the latest. If the programmed loss is inevitable through reasons beyond the contractor's control, then revised rates can be agreed with the client while the revised work is being carried out so that the effect of alterations can be seen by both parties at first hand. This will reduce the amount of paperwork normally involved in recording such revisions.

Figure 28.1 illustrates the pre-cost for the site used in the standard cost example of Chapter 11; the labour values are as illustrated in Figures 11.20–11.27 with the following additions

150 mm diameter concrete pipes	£0.40/m
215 mm brickwork	£7.00/m^2

The pre-cost would normally be calculated on the Friday prior to the week being studied; the number of pay-offs would therefore be known and any new starters would, no doubt, have already been told to report on the following Monday. Thus the last known wage bill can be revised to allow for the current week's pay-offs and new starters and for those during the week being studied. That is

Week 7 wage sheet produced by Thursday of week 8
Pre-cost for week 9 calculated on Friday of week 8
Week 7 wage sheet adjusted for new starters and pay-offs for weeks 8 and 9 is approximately equal to wage sheet for week 9

Pre-costing 251

In the example the assumption has been made that the site will again work a six-day week except for the 22 RB gang, who will work a seven-day week as before. Concrete and mortar mixers will be as week 7.

Like post-costing, pre-costing can be calculated nett or gross. However, as approximations have to be made in pre-costing and no allocation is available, it is generally sufficient to study only measurable work, the normal post-costing system being relied on to indicate any adverse trends

VALUE		PRE-COST				WEEK NO. 9	
MEASURABLE WORK		@	LABOUR, £	@	PLANT, £	@	SUB-LET GROSS L & P, £
Exc. o/s.	20 m^3	1.80	36.00	2.25	45.00		
Exc. bases	50 m^3	2.25	112.50	0.90	45.00		
Exc. drains	20 m^3	2.25	45.00	2.25	45.00		
Exc. basement	800 m^3	0.36	288.00	0.45	360.00		
Blinding	7 m^3	9.00	63.00				
Concrete bases	15 m^3	6.75	101.25				
Concrete floor slab	10 m^3	10.00	100.00				
Concrete drains	3 m^3	8.10	24.30				
Concrete columns	4 m^3	11.70	46.80				
215mm bwk.	20 m^2	7.00	140.00				
150mm dia. pipes	25 m	0.40	10.00				
Cart away	890 m^3					0.90	801.00
			966.85		495.00		801.00
PRELIMINARIES							
LABOUR							
Chainman			45.00				
Unload matls.			20.00				
PLANT							
Conc. mixer					36.00		
Mortar mixer					18.00		
			1031.85		549.00		801.00
+ ON-COSTS 101.27%			1044.95				
GROSS LABOUR			2076.80				
PLANT			549.00				
SUB-LET			801.00				
TOTAL GROSS VALUE			£3426.80				

Figure 28.1 Pre-cost statement for week 9

in overheads. For illustration the following example has been calculated gross.

Programme for week 9

Excavate oversite and cart away	20 m³
Excavate basement and cart away	800 m³
Excavate bases and cart away	50 m³
Excavate drains and cart away	20 m³
Blind bases	4 m³
Blind drains	1 m³
Blind floor slab	2 m³
Concrete bases	15 m³
Concrete floor slab	10 m³
Concrete columns	4 m³
Concrete drain surround	3 m³
215 mm brickwork to m/h	20 m²
150 mm diameter SGW pipes	25 m
Cart away (on lorry)	1100 m³

Cost for week 9

Labour £

Total wages (week 7 for 11 men)	1669
New starters = 2	
i.e. 2 × £1669 / 11	303
Total labour cost (week 9)	1972

Plant

10/7 mixer	36
5/3½ mixer	14
22 RB excavator say 50 h	500
Total plant cost	550

Sub-let

Cart away 1100 m³ @ 70p	770
Total gross cost	£3292

Comparison £

Planned value	3427	
Estimated cost	3292	
Planned gain	£135	or 3.9%

On a site using considerable plant this comparison can be carried out comparing the various element of work.

Labour (gross)

Planned value	2076
Estimated cost	1972
Planned gain	£104

Plant

Planned value	549
Estimated cost	550
Planned loss	£1

Sub-let

Planned value	801
Estimated cost	770
Planned gain	£31

The example shows that although plant and labour-only sub-contractors have been planned economically, domestic labour is planned at a slight loss.

Exercise

(1) Calculate a nett pre-cost, i.e. excluding all variable overheads, for week 10 of the standard cost exercise contract. Assume labour and plant strength to be as week 9 and programme to be as follows

Excavate basement and cart away	600 m^3
Excavate bases and cart away	75 m^3
Excavate drains and cart away	40 m^3
Blind bases	6 m^3
Blind drains	2 m^3
Concrete bases	20 m^3
Concrete floor slab	5 m^3
Concrete drain surround	6 m^3
215 mm brickwork to m/h	30 m^2
150 mm diameter SGW pipes	50 m
Cart away (on lorry)	900 m^3

Chapter 29
Computers

Introduction to computers

Computers are rather like hi-fi sets in the sense that no one part is of much use without the other. They are alike also in the prolific use of jargon, not all of which is helpful to the uninitiated. Just as the radio in a hi-fi set is of no use without programmes being broadcast so a computer is of no use without its programs which are a list of instructions telling the computer what to do and when to do it.

The machine itself is known as *hardware* whereas the programs are referred to as *software*. A compromise term of *firmware* is sometimes used where, for instance, programs have been etched on to microprocessor chips and therefore form part of the machine itself.

Computers – the hardware

The central processor

The heart of the computer hardware is known as the *central processor* which, as its name suggests, carries out all the processing and manipulating of the data fed in. This is where the incredibly small silicon chips or *microprocessors* are contained enabling the computer to deal with enormous amounts of facts and figures in fractions of a second. The central processor is made up of three parts, normally held within a single console.

(1) The *main store* where all the work is done, programs are stored and space is provided for operating the programs and storing the answers.
(2) The *arithmetic unit* which tells the computer how to carry out any mathematics required of it.
(3) The *control unit* which acts as a supervisor issuing instructions and checking that they are correctly carried out.

Input

With similar adaptability to a hi-fi the central processor is capable of being plugged into a variety of input machines, as outlined in Figure 29.1.

Recording is normally via a keyboard, either directly into the central processor or via a variety of storage units such as magnetic or punched paper tapes or cards, cassettes or records. Other possibilities include the use of *readers* such as a *card reader* for understanding cards showing pencilled marks against standard data; *light pens* for marking light-sensitive

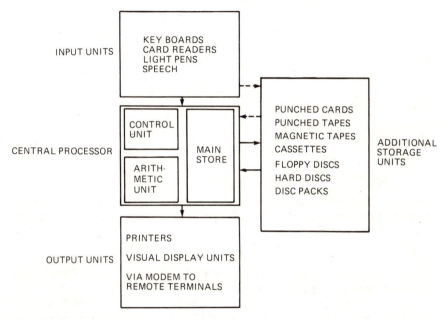

Figure 29.1 Outline of main parts of a computer

screens or for reading off coded numbers such as one sees on food packets and to a limited degree speech which will undoubtedly develop further as techniques advance.

Storage

Any information passing through the central processor can be selected for filing away in the computer's memory or *store*. This may be in the main store part of the central processor or may be in additional storage space. This can be provided by *magnetic tape* either on a spool or in cassettes or in the more easily accessible form of *magnetic discs,* either floppy discs which are similar to the promotional records that one receives through the post from mail order music companies, or hard discs which resemble one or a number of LPs in a dust-free perspex case.

Such discs range in price from small single-sided floppy discs at only a few pounds each, up to multiple packs of hard discs costing several thousand pounds. The difference in cost reflects the varying storage capacities. Just as LPs play longer than 45s on a hi-fi, so hard discs will long outlast a floppy both in storage and wearing capabilities.

Output

The output from the processor can be via a *printer*, i.e. typed by the computer on to sheets or rolls of paper; displayed on a television type screen, *visual display unit* (VDU), recorded for later use on any of the storage media previously mentioned or held for a short while in the central processor until required. By connection to a MODEM (all British Telecom exchanges are to be converted from analogue to digital switching by 1992 which will cut out the need for a MODEM), i.e. the link to the present telephone network the data can be sent to a *terminal*, i.e. a remote unit in a different part of the building, country or indeed anywhere in the world that has access to a suitable telephone line.

Such terminals are usually just a keyboard and VDU or printer or they may be another computer when they are termed an *intelligent terminal*.

Batch or interactive processing

The various methods of input of data into a computer divide clearly into two types and provide two distinct systems of processing data.

(1) Batch processing
(2) Interactive

Batch processing

Batch processing is when all the data to be processed by a particular program are collected and passed to the computer *en bloc*, either on punched cards or some other form of input. This form of processing is suitable for straightforward facts and figures that do not required interpretation, e.g. payroll, invoices, etc.

Interactive

Interactive is when the input of data is entered directly via a keyboard into the computer producing an immediate result on the VDU and prompting the user to scrutinise the data as they are processed. This is especially useful for interpretation of cost control data where different terms may have been used for the same item of work, e.g. formwork and shuttering, cills and sills, etc., which the computer will not recognise as being the same. Correction can easily be made by the user, before a print-out is made or the data are filed away on storage discs. Unreasonable or illogical answers can be questioned and corrected at source and items grouped together or kept separate at the will of the user.

Mainframes and microcomputers

Although costs of hardware have fallen considerably over the years there are still vast differences in price reflecting the relative facilities, capacities or speed that each machine offers. At the bottom end of the market are the home microcomputers such as the Sinclair ZX 81, which utilises normal TV screen and cassette recorder, through the small business microcomputers such as the Commodore Pet, Tandy and Apple which come complete with printer and drive units for floppy discs. There are also more expensive minicomputers capable of running more than one terminal at a time, i.e. a number of VDUs, printers, etc., and then upwards in cost to the more powerful multi-unit mainframe computers having vast on-line storage, i.e. immediately accessible memory, a variety of terminals, readers, printers, etc.

Bits and bytes

The internationally accepted system for mathematics is the decimal system, i.e. to count from 0 to 9 and then use combinations of these digits for all higher numbers. A computer however generally uses a binary system, i.e. it counts from 0 to 1 and then starts to use combinations of these two digits for numbers higher than 1, e.g. 2 is represented by 10 in binary 3 by 11, 4 by 100 and so on. The computer uses binary because it retains its information by electronic signals that are either off or on, i.e. 0 or 1.

Each of these digits whether 0 or 1 is known as a *bit*. By grouping together a number of bits not only can the decimal numbers 0–9 be represented but also the letters of the alphabet, punctuation marks, mathematical symbols, etc. A group of 8 bits or *bi*nary dig*its* for instance produces a total of 2^8, i.e. 256 combinations. This group of bits is known as a *byte* and is usually 8 bits long though the more powerful minis and main-frame computers may use 16 or even 32 bit bytes.

The memory of a computer and storage capacity of discs are measured by the maximum number of bytes that they can hold at any one time.

1000 bytes is shortened to 1 K
In precise terms 1 K = 2^{10} which is 1024 bytes
1 000 000 bytes is known as 1 megabyte

Thus a computer will be specified as having 4 K, 32 K, 64 K, etc., of memory (but see notes on ROM and RAM).

For storage purposes a single-sided 5¼ in diameter floppy disc may be specified as capable of storing 500 K whilst a Winchester disc, which is a single hard disc, may have 10 megabytes of store available. These figures vary depending on the type of disc and machine that uses them.

ROM and RAM

Although a computer may have, say 48 K of memory not all of that power is available for the user. Some of the memory is taken up by the computer itself in the instructions built-in by the manufacturer to tell it how to operate. This built-in program is known as *systems software* and the amount of memory it uses is known as ROM, i.e. *ready-only memory*, this may amount to 16 K of the original 48 K available. The users programming is known as *applications software* and the memory remaining for this as RAM, i.e. *Random Access Memory*. In the example above 32 K of RAM would be left available to be used time and time again, particular programs and data being stored outside this 32 K of the central processor when not in use.

The relative ease and cheapness with which memory capacity can be increased in modern computers has helped in making those computers easier to use. It is of little concern to the user to lose, say one-third of the capacity of a computer to the systems software if that systems software gives him an easier-to-use computer language and therefore makes it simpler for him to produce his own programs on the still substantial 32 000 bytes of memory that remain.

Computer language

The binary codes mentioned previously are in a form that the computer can easily interpret. This is referred to as *machine code*. To produce lengthy programs in this way, however, is a laborious task and short cuts have been developed so that when simple words or symbols are entered the computer automatically looks them up on its *translator* or dictionary and converts them into machine code. These key words which are not immediately recognised by the computer constitute what is known as *low level languages* because they are at a level pretty near the computer code itself. Examples of this are found in languages such as PLAN (Programming LAnguage Nineteen hundred) the assembly language of the ICL 1900 series computers and in other assembler languages perculiar to each make of machine.

The writing of instructions or programs in these low-level languages is still a cumbersome task, however, and has developed a stage further to encompass whole lists of instructions within one work or symbol. These languages are further removed from the original machine code and are therefore known as *high-level languages*. They need more than just a dictionary for translation, they require what may be described as a grammar book which is known as the computers' *compiler*. The compiler is able to interpret messages received in a high-level language and translate them into machine code.

The high-level languages take a number of different forms depending on why they were developed. Some of the more usual examples are FORTRAN (FORmula TRANslator) and ALGOL (ALGOrithmic Language) developed for scientific and mathematical use; COBOL (COmmon Business Oriented Language) for business and commercial applications and BASIC (Beginners All-purpose Symbolic Instruction Code) developed initially as a simple language for beginners but rapidly becoming accepted as the language of micro-computers with the increase in DIY programming. (An example of a simple program written in BASIC is shown on page 262.)

The use of so many high-level languages including variations within the same language make it difficult to use programs written for one computer on a different model. As the range of models of computer grows so the variations in high-level languages grow. Many of these models, however, are based on the same basic microprocessor chips. An attempt at solving the language problem has therefore been developed known as CP/M which is an operating system used by some hardware and software companies that allows the user to work in a high-level language but operates the computer at low-level in a form interchangeable between models.

Such interchangeability has obvious advantages; as hardware companies change their models or even go out of business it is essential to have some thread of continuity that CP/M offers. In addition to this many companies offer deals in updating their equipment and programs as new applications of computer technology develop.

Further information on computers can be obtained from the following

CICA – The Construction Industry Computing Association, Guildhall Place, Cambridge CB2 3QQ. Telephone (0223) 311246 (previously known as the Design Office Consortium).

NCC – The National Computing Centre Limited, Oxford Road, Manchester M1 7ED. Amongst their publications is a cassette and booklet entitled *How to Choose your Small Business Computer*. The NCC also runs a Microsystems Centre set up to provide training, information and advice to potential users of microcomputers at 11 New Fetter Lane, London EC4A 1PU. Telephone 01 353 0013/4/5.

Building Advisory Service, 18 Mansfield Street, London W1M 9FG. Telephone 01 636 2862. The BAS runs appreciation courses on computers geared to the construction industry.

MAP – Information Centre, Dean Bradley House, 52 Horseferry Road, London SW1P 2AG. Telephone 01 212 3411/4. MAP is the Department of Industry Programme to Encourage the Application of Microelectronics. A diary of forthcoming conferences/seminars/courses/exhibitions is issued periodically from the above address at Telephone 01 212 3415.

Office Equipment News, 75–77 Ashgrove Road, Ashley Down, Bristol BS7 9BR. The OEN issues a free monthly newspaper featuring office equipment including the latest models of computers and software packages.

Chapter 30

Computer costing

The software

How do we create the computer program to produce the various cost reports, analyses, cross references, summaries, etc., that are required from any cost system? A number of alternatives exist.

  (1) Write a made-to-measure program.
  (2) Buy a ready-made program and alter it to suit your requirements.
  (3) Buy a program package that will assist you in setting up the system for yourself.

Made-to-measure program

In order to convert data put into a computer into the format, order and details that are required to be printed out, recorded, accumulated or erased, a program needs to be tailor-made to suit the particular job of work that is being requested. This program or software is the most difficult part of setting up a computer-operated system. Made-to-measure programs can, however, be produced by any combination of the following

  (1) by a software house. It is undoubtedly an attractive proposition to take advantage of the expertise that is available in reputable software companies. However, the development costs of producing a program can frequently be higher than the cost of the hardware. A typical weekly standard cost system as described in Chapter 11 would cost several thousand pounds to develop and subsequent developments, revisions or amendments to the program would entail further expense. However, provided a company is able to define exactly what it wants from a system, then it is worthwhile obtaining quotations for such a program
  (2) by your own company. You may already have a computer or data processing department in which case they may be able to help in producing a program to suit your needs or at least take data already processed by a computer and re-sort them into a form that is useful

PAYROLL TRADE ANALYSIS — THIS WEEK – COST PER WORKING HOUR — CONTRACT EXAMPLE — WEEK ENDING 28/3/82

TRADE GROUP	JOINERS	BRICKLAYERS	STEELFIXERS	SCAFFOLDERS	MECHANICS	DRIVERS	LABOURERS
BASIC WAGES	1.93	1.85	1.91	1.89	2.01	1.80	1.65
EXCESS OVERTIME	0.04		0.05	0.03	0.03	0.07	0.03
INCLEMENT WEATHER	0.02		0.03	0.01			
BONUS	0.90	1.15	0.85	0.80	0.52	0.45	0.40
TRAVEL TIME	0.08	0.08	0.07	0.03	0.13	0.06	0.03
FARES	0.03	0.05	0.03	0.01			0.02
SUBSISTENCE							
TOOL MONEY AND EXTRAS	0.08	0.05	0.05	0.05	0.10	0.07	0.04
SICK PAY	0.01				0.07		0.01
PUBLIC HOLIDAY WITH PAY							
EMPLOYER'S GNI	0.42	0.43	0.40	0.39	0.39	0.34	0.30
HWP	0.25	0.26	0.26	0.25	0.28	0.22	0.24
TOTAL COST	3.76	3.87	3.65	3.46	3.53	3.01	2.72
HOURS WORKED	839	178	210	365	32	161	622
NUMBER OF MEN	23	5	6	10	1	4	17
AVERAGE HOURS WORKED/MAN	36	35	35	36	32	40	37

Figure 30.1 Analysis of payroll by trades

for cost control purposes as opposed to the more usual accountancy requirements.

For instance, it is useful for both cost control and estimating purposes to know the weekly and periodic costs of labour trade by trade and the breakdown of that cost as suggested in Figure 30.1. This at least goes some way towards taking the drudgery out of cost calculations and assists in more accurate pricing of items in different trades. In the example illustrated, all workers in a particular trade have been included under that trade heading including chargehands, tradesmen, tradesmens' labourers and apprentices

(3) DIY – do-it-yourself. Numerous programming courses exist at local colleges, at specialist training establishments and at some of the hardware outlets such as the courses run by TANDY and COMMODORE computer centres.

A simple *listing* of a program written in the language BASIC is shown in Figure 30.2. This example illustrates how the computer can be programmed to ask a question and then to use the answer in asking further

```
10  PRINT"       EXERCISE IN BASIC"
20  PRINT"CALCULATION OF OUTPUTS AND UNIT COSTS"
30  PRINT "-------------------------------------------------- "
40  PRINT
50  PRINT" PLEASE ENTER THE COST HEADING"
60  INPUT A$
70  PRINT" ENTER THE HOURS SPENT ON "A$
80  INPUT H
90  PRINT" ENTER THE COST PER HOUR ON "A$
100 INPUT R
110 PRINT" ENTER THE MEASURE OF WORK DONE ON "A$" AND THE UNIT OF MEASUREMEN"
120 INPUT M,T$
125 REM C = TOTAL COST  O = OUTPUT  U = UNIT COST
126 REM H = HOURS  R = COST/HOUR  M = MEASURE
127 REM A$ = COST CENTRE  T$ = UNIT OF MEASUREMENT
130 LET C=H★R
140 LET O=H/M
150 LET U=C/M
160 PRINT A$" OUTPUT IS "O" HOURS PER "T$" AND UNIT COST IS U" PER "T$
170 IF O < 0.5 GOTO 250
180 IF O > 1.5 GOTO 270
190 PRINT" PROGRAM COMPLETE"
200 END
250 PRINT" OUTPUT FOR "A$" LOOKS UNUSUALLY GOOD"
260 GOTO 280
270 PRINT" OUTPUT FOR "A$" LOOKS UNUSUALLY POOR"
280 PRINT" PLEASE CHECK THAT MEASURE AND HOURS ARE CORRECT"
290 GOTO 190
```

Figure 30.2 Example of simple computer program written in BASIC language

questions. It can be instructed to look for unusual figures and then warn the operator if results look incorrect.

Following through this program step by step we can see that lines 10 and 20 print a heading on the screen whilst line 30 underlines this heading. Line 40 leaves a space before line 50 states 'Please enter the cost heading'. Anything written between quotation marks after a PRINT instruction will appear on the screen when that line of the program is reached including the spaces shown on line 10 to centralise the heading.

EXERCISE IN BASIC

CALCULATION OF OUTPUTS AND UNIT COSTS

PLEASE ENTER THE COST HEADING
PLACING CONCRETE
ENTER THE HOURS SPENT ON PLACING CONCRETE
100
ENTER THE COST PER HOUR ON PLACING CONCRETE
5.00
ENTER THE MEASURE OF WORK DONE ON PLACING CONCRETE AND THE UNIT OF MEASUREMENT
25 CU.M.
PLACING CONCRETE OUTPUT IS 4 HOURS PER CU.M. AND UNIT COST IS £20 PER CU.M.
OUTPUT FOR PLACING CONCRETE LOOKS UNUSUALLY POOR
PLEASE CHECK THAT MEASURE AND HOURS ARE CORRECT
PROGRAME COMPLETE

Figure 30.3 Print-out from program shown in Figure 30.2

At line 60 the computer waits for the cost heading to be entered and records this against its reference A. The $ sign simply tells the computer to expect a *string* of letters or words for A and not to expect a number. Line 70 now prints the next question using whatever string of letters was entered for A $ as part of the question. Line 80 waits for the hours to be entered against the reference H. Similarly lines 90 and 100 ask and receive the rate per hour referenced R. All of these questions could have been asked together as illustrated in lines 110 and 120 where a mixture of numbers and words are dealt with in the same line.

The reason lines are numbered in 10s is to allow insertions as those shown from 125–127. These particular insertions have been put in to remind the programmer what all the different letters are meant to represent. REM is simply short for *remark*.

Lines 130–150 lay down any mathematical formula required to calculate the results which are printed by line 160; the symbol ★ being used for multiplication and / for division. Before finishing the computer has been programmed to carry out a couple of checks. In line 170 if the output is less than 0.5 it will go to line 250 and then 280 and print a warning. If, however, it gets past line 170 it will then be further checked in line 180 to see if the output is greater than 1.5 which will lead it to the warning in line 270 and then 280. Finally all routes lead to line 190 with a message that the program is complete and line 200 will instruct the computer to this effect. Figure 30.3 shows what happens when this program is run on the computer. The shaded areas indicate an input from the keyboard, the remaining text being produced almost instantaneously by the program.

Such a program as the above could be written after only a short period of study but like the learning of a foreign language fluency cannot be obtained so easily and much practice is needed before lengthy, complicated programs can be mastered.

Ready-made programs

There are a number of ready-made programs on the market which produce cost data of various kinds and a number of construction companies have developed their own software to suit their particular needs. However, herein lies the problem. No two companies will require identical sets of data presented in exactly the same format and to the same degree of detail. Therefore in any ready-made package either the company has to accept the straightjacket of the system as it stands or be involved in the expense of reprogramming the package to suit its own requirements.

As with made-to-measure programs, a software house will be only too willing to reshape the package to fit, the company computer department may be able to help or with the majority of the donkey work done by the package and some knowledge of programming it may be possible to make amendments oneself.

At contract account level a number of ready-made programs exist for the small and medium-sized builder that require very little tailoring depending of course on a company's individual accounting demands. Advice on such systems can be obtained from any of the addresses shown on page 259.

Program packages

With the increased use of DIY computing a number of programming aids have been coming on the market that make it possible for the user to create his own programs with no knowledge of computer languages whatsoever. This will undoubtedly develop further and will ease the burden of writing lengthy programs for numerous common applications of a computer.

There are four fields of application in which we are particularly concerned in relation to cost control in the construction industry.

(1) *Number crunching*. The calculating, abstracting, accumulating, etc., of figures is merely a series of different uses of arithmetic, trigonometric and algebraic functions. Program packages such as VISI-CALC (obtained from most computer outlets in the UK) recognise that many problems are commonly solved with a calculator, a pencil and a sheet of paper. VISICALC combines the convenience and familiarity of a pocket calculator with the powerful memory and electronic screen capabilities of a computer. The screen becomes a rough book. A print-out being available at any time of all or part of

the rough book calculations, or work can be stored electronically at any time for future reuse or reference.

The program consists of a gigantic chart 63 columns wide headed from A to BK and 254 lines deep numbered from 1 to 254. Thus each of the 16 002 squares on this chart has a unique reference commencing at A1 and ending at BK 254. Not all of these squares can be used at the same time, however; they are merely referenced to allow the choice of working across the chart or down the chart. In each of these squares either labels or values can be entered.

Labels consist of words or titles such as *total cost* or *JCB 3CX EXCAVATOR*. Values consist of numbers or formulae, e.g. 168.35 or A3 + B7/D4 − C5 ★ F8, the / sign being the computer symbol for divide and the ★ sign for multiply.

It takes about a day of self-teaching to fully understand the intricacies of VISICALC; how to split the screen into windows in order to see different parts of the chart at the same time, how to format the words and numbers in each box, how to set column widths, move, repeat, insert or delete entries, how to store data on disc and recover that data when required. Once the mechanics of operating the system have been mastered, however, there are innumerable applications on a construction site that lend themselves to the use of this means of calculation.

Figure 30.4 illustrates the use of VISICALC for producing trade costs and shows how each square in the chart is set up, e.g. B8 is labelled *formwork* and D2 to J2 are labelled with dashes to produce a dotted line. Other squares have been given a value, e.g. F5 = D5 × E5 and D10 is the sum of D5 and D8 whilst other squares have been set at zero to indicate that data are required.

Figure 30.5 continues this theme by producing the to-date calculations by adding together the this week figures and the previous to-date totals which the computer can read electronically from data stored on a floppy disc the previous week. When figures are entered against the appropriate cost headings the computer processes the various calculations almost instantaneously and a print-out can then be produced as shown in Figure 30.6. The previous week's cumulative figures are shown for reference in Figure 30.7 but these would not normally need to be printed out.

There are, however, a couple of disadvantages to this type of program. The major drawback being one of size. The example illustrated in Figures 30.3–30.7 which was run on a microcomputer with a 32 K capacity ran out of memory at about 30 cost headings, VISICALC having only 10 K of memory available on this size of hardware, i.e. 10 000 bytes of information. This problem is not insurmountable, however, as the headings can be sectionalised, e.g. by trade or by areas of a contract but operation then becomes somewhat cumbersome if input data is not sub-divided in strictly the same sections and requires more posting of data backwards and forwards between the computer and the floppy discs. A second problem is that in entering data it is all too easy to accidentally erase or alter the formulae thus rendering the results open to error.

	a	b	c	d	e	f	g	h	i	j	k
1			THIS WEEK								THIS WEEK
2		trade	measure	hours	lab rate	lab cost	plant	total	output	unit cost	measured per
3											
4											
5		mix conc	0.00	0.00	0.00	d5 x e5	0.00	f5 + g5	d5 / c5	h5 / c5	cu. metre
6		place conc	0.00	0.00	0.00	d6 x e6	0.00	f6 + g6	d6 / c6	h6 / c6	cu. metre
7		fix reinf	0.00	0.00	0.00	d7 x e7	0.00	f7 + g7	d7 / c7	h7 / c7	tonne
8		formwork	0.00	0.00	0.00	d8 x e8	0.00	f8 + g8	d8 / c8	h8 / c8	sq. metre
9											
10		TOTALS	sum c5..c8	sum d5..d8		sum f5..f8	sum g5..g8	sum h5..h8			
11											

Figure 30.4 Program to produce week's trade cost set up using VISICALC

	l	m	n	o	p	q	r	s	t	u
1		TO DATE								TO DATE
2										
3	trade	measure	hours		lab cost	plant	total	output	unit cost	measured per
4	---	---	---		---	---	---	---	---	---
5	mix conc	c5 + w5	d5 + x5		f5 + z5	g5 + aa5	p5 + q5	n5 / m5	r5 / m5	cu. metre
6	place conc	c6 + w6	d6 + x6		f6 + z6	g6 + aa6	p6 + q6	n6 / m6	r6 / m6	cu. metre
7	fix reinf	c7 + w7	d7 + x7		f7 + z7	g7 + aa7	p7 + q7	n7 / m7	r7 / m7	tonne
8	formwork	c8 + w8	d8 + x8		f8 + z8	g8 + aa8	p8 + q8	n8 / m8	r8 / m8	sq. metre
9		---	---		---	---	---			
10		sum m5..m8	sum n5..n8		sum p5..p8	sum q5..q8	sum r5..r8			
11		---	---		---	---	---			

	w	x	y	z	aa
1	CUM B/F				CUM B/F
2					
3	measure	hours		lab cost	plant
4	---	---		---	---
5	0	0		0	0
6	0	0		0	0
7	0	0		0	0
8	0	0		0	0

Figure 30.5 Program to accumulate trade costs

	THIS WEEK						THIS WEEK	
trade	measure	hours	lab rate	lab cost	plant	total	output	unit cost measured per
mix conc	500.00	200.00	5.00	1000.00	1500.00	2500.00	0.40	5.00 cu. metre
place conc	500.00	800.00	4.50	3600.00	1400.00	5000.00	1.60	10.00 cu. metre
fix reinf	10.00	200.00	5.00	1000.00	50.00	1050.00	20.00	105.00 tonne
formwork	800.00	1500.00	5.00	7500.00	500.00	8000.00	1.88	10.00 sq. metre
TOTALS	1810.00	2700.00		13100.00	3450.00	16550.00		

	TO DATE						TO DATE	
trade	measure	hours		lab cost	plant	total	output	unit cost measured per
mix conc	15500.00	7200.00		35000.00	41500.00	76500.00	0.46	4.94 cu. metre
place conc	15500.00	20800.00		87600.00	31400.00	119000.00	1.34	7.68 cu. metre
fix reinf	110.00	1700.00		8000.00	350.00	8350.00	15.45	75.91 tonne
formwork	6800.00	14000.00		72500.00	4500.00	77000.00	2.06	11.32 sq. metre
	37910.00	43700.00		203100.00	77750.00	280850.00		

Figure 30.6 Print-out of trade cost from program illustrated in Figures 30.4 and 30.5

CUM B/F measure	hours	lab cost	CUM B/F plant
15000	7000	34000	40000
15000	20000	84000	30000
100	1500	7000	300
6000	12500	65000	4000

Figure 30.7 Previous week's cumulative figures stored for reference on floppy disk

The program is nevertheless an exceptionally useful aid and can be used for numerous construction calculations such as, for instance, those illustrated in the assessment of earthmoving costs in Chapter 20 which can take a crippling amount of time to produce manually because of the numerous variables involved.

This task can be set up using VISICALC in only a couple of hours and then is stored on floppy disc for use with any combination of plant hire rates, fuel consumption and fuel costs per litre, haul speeds and lengths, combinations of machines per team, working hours per week, etc. An example of such a usage is illustrated in Figure 30.8 where an alteration to any of the shaded figures would initiate an immediate recalculation of the remaining cost and output results.

(2) *Data base systems.* Computers are exceptionally efficient at sorting through data and abstracting the appropriate records. Many data base systems are available for use on various micro-computers and can be classified basically into two different types.

(i) Those requiring the disk containing the software package to be present whenever the program is being run.
Some examples of this are
DMS – by Compsoft Limited, Great Tangley Manor Farm, Wonersh near Guildford, Surrey GU5 0PT. For use on Commodore computers or any micro using CP/M operating system.
Silicon Office – which includes word processing and other additional features by The Bristol Software Factory, Kingsons House, Grove Avenue, Queen Square, Bristol, BS1 4QY. For use on Commodore 8096 or 8032 upgraded to 96K.
Administrator – by Stage One computers, 300 Ashley Road, Parkstone, Poole, Dorset.
(ii) Program Generators. Any program created for a particular application is stored on a floppy disc and can be run quite independently of the program generator disc. Any number of copies of the generated programs can be made for use at different locations whilst the generator disc is still being used to create yet more programs for other applications. The programs are generated in the language and dialect understood by the computer being used and can subsequently be edited or changed in the normal way if amendments to the generated program are required.

```
CALCULATION              OF              EARTHMOVING                 COSTS
*************************************************************************
Calculations are based on the following rates :-
------------------
     CAT 627 Scraper
------------------

          Plant hire          1000.00     hire per week
       Spares & cons           100.00     cost per week
    Fuel consumption              100     litres/wkg hr
------------------
     CAT D8 Dozer
------------------

          Plant hire           700.00     hire per week
       Spares & cons           100.00     cost per week
    Fuel consumption               50     litres/wkg hr
------------------
     CAT 16 Grader
------------------

          Plant hire           800.00     hire per week
       Spares & cons           250.00     cost per week
    Fuel consumption               40     litres/wkg hr
------------------
     General costs
------------------

             Drivers             5.00     per attend hr
                Fuel             0.15           per litre
             Payload            11.00     cu metres/load
          Haul speed              350     m/standard min
        Return speed              250     m/standard min
        Load & tip time          2.50standard mins/load
------------------
     Working hours
------------------

    Attendance hours               60     hours/week
           % working              75%
     = working hours               45     hours/week
```

```
----------------------------------------------------------------
BUILD    UP    OF    COST    PER    MACHINE    PER    WORKING    HOUR
----------------------------------------------------------------
                     CAT 627 scraper      CAT D8 Dozer      CAT 16 Grader
                     ---------------      ------------      -------------
        Plant hire            22.22             15.56              17.78
     Spares & cons             2.22              2.22               5.56
              Fuel            15.00              7.50               6.00
            Driver             6.67              6.67               6.67
                             ------            ------             ------
TOTAL/WORKING HOUR            46.11             31.94              36.00
                             ------            ------             ------

----------------------------------------------------------------
BUILD    UP    OF    COST    OF    TEAM    PER    WORKING    HOUR
----------------------------------------------------------------
            3      number    CAT 627 scrapers              138.33
            2      number    CAT D8 dozers                  63.89
            1      number    CAT 16 grader                  36.00
          600 pounds cost of :-bowsers,fitter etc           13.33
                                                           ------
TOTAL  TEAM  COST    PER             WORKING    HOUR       251.56
                                                           ------
****************************************************************
     RESULTS BASED ON           500          Metres          Haul
----------------------------------------------------------------
              Total cycle time  in standard mins            5.36
              Loads    per     standard    hour            11.20
              Cubic metres /hour of one machine           123.20
              Cubic metres /hour of above team            369.60
              ------------------------------------        ------
              COST    PER    CUBIC     METRE                0.68
              ------------------------------------        ------
****************************************************************
```

Figure 30.8 Calculation of earthmoving costs using VISICALC

Some examples of this are

Nucleus – by Compact Accounting Services Limited, Cape House, Cape Place, Dorking, Surrey. For use on Adds, Rank Xerox or 64K micros using CP/M operating system.

C.O.R.P. – by Dynatech, USA, available through Apple dealers.

The Last One – by D.J. 'Al' Systems Limited, Station Road, Ilminster, Somerset TA19 9BT.

Discovery Sys–Gen – by Software Consultants, 40 Triton Square, London NW1 3HG. For use on Compucorp or Jacard computers or 64K micros using CP/M operating system.

Common to all of these aids is the adoption of filing cabinet terminology which, although simple, can create great confusion if the terms are used loosely.

(i) System. The particular draw of a filing cabinet. Just as a filing cabinet may contain a number of drawers so also may a disc contain a number of systems all separately labelled, e.g. 'Weekly Cost System', 'Personnel System', etc.

(ii) File. The separate files within the drawer. A *master file* is for reference purposes only, e.g. 'list of plant rates', whereas a *transaction file* is for recording repetitive data, e.g. 'record of hours concreting bases each week', etc.

(iii) Field. Each file will contain its own standard list of headings such as 'week ending', 'type of plant', etc. The key field is the main reference field on that file, e.g. task ref., etc.

(iv) Record. Up to this point all that has been described is a series of blank forms in a filing cabinet, the term *record* is used to denote one of these files with some data entered against any of its fields.

In setting up a weekly costing system, the first task is to decide how the files are going to be arranged. It is possible to enter all data on to the same file format using field names such as

(1) Week ending
(2) Gang name
(3) Location of work
(4) Trade
(5) Description of task
(6) Number of man hours on task
(7) Cost per man hour
(8) Plant costs involved
(9) Measure of work done
(10) Unit of measure
(11) Value per unit of measure
(12) Comments

However, it may be preferable to enter plant expenditure in more detail and on a separate file to the labour allocation. A master file could also be set up to look up the unit values of each task rather than have to enter this through the keyboard each time that task occurred. The unit of measure could similarly be looked up, as could the cost per man hour for the particular trade or particular gang.

Having decided on the fields in each of the files, the computer is ready to accept data to form a series of electronically stored records. Whenever required these records can be sorted to produce a report in whatever combination of selection criteria has been requested, e.g. list all records against field 5 *description of task* together with fields 9, 10 and 6, i.e. *measure, unit* and *man hours* carried out by Smith's gang (in field 2) in the boiler house (in field 3). Calculations

```
NO OF MEN    COST / HR COST PER HOUR ABOVE 4.00  * = 10p CONTRACT

    12       4.20 **                             Browns Factory

     3       4.55 *****                          Devon dairy

     5       4.82 ********                       Swindon offices

    10       5.00 **********                     Cambridge centre

    14       5.30 *************                  Jones factory

     4       5.31 *************                  Birmingham school

     8       5.35 *************                  Stoke road

    16       5.40 **************                 Derby offices

    22       6.00 ********************           Leeds bank

     5       6.74 **************************     Watford centre
```

Figure 30.9 Simple graphics printed by computer

can be incorporated in the report to produce unit costs, outputs, gains and losses, etc. and totals and sub-totals produced where required.

With some DMS software the selections can even be made to scan a sentence for key words, e.g. to look for any reference to B works under *location* or any reference to delay or disruption under *comments*, any item with a *plant cost* of over £100 or any trade showing a *gain* of less than 5%, etc.

As with a manual system great care must be taken in setting up the list of names to be used for each file and each field. Advice must be sought from the software supplier on the number of records that will be possible on a particular computer without having to change storage discs. This may not be too bad if each disc can hold sufficient data for, say three months' work. However, it would be very cumbersome if each week required a separate storage disc and a selection of data was required from the whole period of a two-year contract involving over 100 floppy discs. The number of discs required will depend on the density of the discs used and on the degree of detail being recorded.

(3) *Word processing.* Many packages now exist as aids in the preparation of reports, letters, circulars, etc. These enable standard preambles, paragraphs or whole sections of a document to be stored electronically for insertion at the appropriate point in any cost report or summary.

Words, lines, paragraphs, etc., can then be inserted or deleted on the computer screen and relevant figures from other packages transferred into the report before finally printing for distribution. Such aids as word processing can save days, even weeks, in the preparation of lengthy reports which may mean the difference between initiating corrective action or simply holding a post-mortem.

(4) *Graphics.* Most computers are capable of simple graphics in as much as asterisks can be printed to represent a figure or a range of figures and thus produce a histogram or bar graph. In the example illustrated in Figure 30.9 the average labour rates of joiners on various sites have been sorted in ascending order and printed out as a histogram where each 10p above £4.00/h is represented by an asterisk.

To produce this data as a proper graph would require much more sophisticated programming and additional hardware in the form of a graphical plotter which is able to plot any number of lines on a chart complete with any annotation required.

Time studies by microprocessor

The use of microprocessor technology is showing its application in the field of work study. This appears to be developing along the following lines

(1) *Computer programs.* A number of companies and consultants have produced micro-computer programs which carry out the numerous calculations, sorting and cataloguing of data inevitably involved in a detailed time study. Time spent in writing up the study is therefore reduced, making it possible to produce study results more quickly, more accurately and more frequently than is possible by conventional means. Time is lost, however, in the entering of data into the computer after the physical study has been completed.

(2) *Data collectors.* This delay can be partly overcome by using recording equipment on site that can be electronically fed into a computer for subsequent analysis without the need to enter the data via a computer keyboard. An example of such recording equipment is the Datamyte 1000.* Connecting cable battery charger, programs and

---

* More information on the Datamyte series can be obtained from Structural Monitoring Limited, 25 Blythswood Square, Glasgow G2 4BL.

computer are not included in this cost. A futher example is the TS 20 Time Study Board manufactured by OPTECH Automation Limited, Units 12/13, Loomer Road Industrial Estate, Stoke-on-Trent, Staffs.

(3) *Data collector/processor*. An extension of this concept is to have a portable data recording keyboard that will process its own calculations without connections to a computer. A data collector/processor such as the Datamyte 803 will do just that, permitting observed ratings to be recorded as the elements occur in a range from 55 to 150 rating on up to 20 work elements.

Exercises

(1) Using a microcomputer and VISICALC package enter the labels and values illustrated in Figures 30.4 and 30.5. Save your program, i.e. command/SS under an appropriate title and then input imaginary data at each of the squares that have a 0.00 entry. When you have mastered this stage save the data, i.e. command/S#S between co-ordinates M5 and Q8 under a heading of Week number 1. Clear week 1's data from the screen and enter some different imaginary data for week 2. Finally bring the previous cumulative forward by loading, i.e. command/S#L Week number 1 between the co-ordinates W5 and AA8. The to-date figures will then automatically accumulate.

(2) Using a microcomputer and VISICALC package set up a program that will produce the calculations shown in Figure 30.8 and then vary any of the highlighted data to calculate the different costs per cubic metre.

(3) Chapter 23 shows how the average cost per man hour can be calculated for different hours worked per week. The example illustrated in Figures 23.1 and 23.5 is based, however, on one hypothetical situation. Using a microcomputer and VISICALC package set up a program to calculate the average cost per man hour based on any number of variances to this situation, e.g. different proportion of men on subsistence, change in level of Graduated National Insurance, any average number of kilometres between men's homes and site, etc.

Index

Absenteeism, 58, 81, 129, 211, 218, 223, 236
Accounts
   contract, 3, 4–11
   firm's, 3
   profit and loss, 3
Action on losses, 141–149
Action sheet, 146
Activity sampling, 161, 167, 223
Adjustments to costs, 71, 126
Allocation sheets, 217
   coded descriptions, 61
   completed by ganger, 59
   completed by other than ganger, 63
   daily time, 89–96
   overheads, 66
   overtime, 66
   plant, 63–65
   standard descriptions, 61
Allocation time, 63
Allowed time, 70, 178, 185, 191, 193, 233, 234, 235
Apprentices, 126
Assessed bonus, 237
Attendance
   on sub-contractors, 19
   on tradesmen, 70, 123

Backfilling, 28
Balance, 56, 112–113
Bank charges, 3
Barrow runs, 58, 130
Basic time, 154, 185
Batching plant, 205
BATJIC agreement, 239
Bench marks, 166
Bill of Quantities, 10, 33, 72, 121, 228
Blockwork, 29, 34
Bonus
   assessed, 237
   costing, 82, 228
   fixed, 237

Bonus (*cont.*)
   payments, 86, 122, 216
   schemes *see* Incentives
   shares, 99, 100
   spot, 237
   targets, 88
Branch drains, 26
Breakdown of estimator's rates, 72
Brickwork, 29
British Standard on Time Study Techniques of Work Measurements, 141, 150
Budget, 33, 80, 108, 208
Bulking, 35
Buying margin, 5, 13

Carcassing, 29, 31
Cart away, 101, 103, 173
Cash records, advantages and disadvantages of, 86
Cats eyes, 24
Cement, 146
   stock, 146
Chainboy, 81
Charts, 127, 178
CITB levy, 211, 217, 219, 220, 234
Civil engineering, 56
Clean office, 81
Clean public roads, 123
Clean site, 21, 123
Client's effect, 72
Clocking on, 89, 225
Coded descriptions, 61
Cofferdams, 28
Computers, 62, 72, 144, 181, 254
   bits and bytes, 257
   costing, 260
   hardware, 254
      batch processor, 254
      central processor, 254
      input, 254
      output, 256

278 Index

Computers (cont.)
  hardware (cont.)
    storage, 255
  language, 258
  made-to-measure program, 260
  mainframes and microcomputers, 256
  program packages, 264
  ready-made programs, 264
  ROM and RAM, 257
  software, 260
  time studies, 274
Concrete, 103
  delivery, 173
  mixing, 123, 129, 146, 203, 208
  over-cementing, 145
  placing, 173
  pouring, 238
  ready-mixed, 203
  site-mixed, 203
Consumables, 86, 148
Contentious items, 8
Contentious payments, 8
Contingency allowance, 157
Contract accounts, 3, 4–11
Contracts
  earthmoving, 56
  housing, 33
  motorway, 56
Cost(s) and costing
  adjustments, 126
  basic, 50
  in cash, 86
  civil engineering, 56
  comparisons, 142
  by computer, 144
  fixed, 205, 210
  fuel, 20, 227
  gross, 50
  gross cash, 41
  in hours, 86
  improvement, 141
  labour see Labour cost
  marginal see Marginal costing
  nett, 41
  per man hour, 148
  plant, 5 see also Plant
  prime see Prime costs
  site, 3
  of small items, 120
  spot, 63, 130, 145
  by stages, 124
  standard see Standard costs and costing
  statement, 55, 86
  summary, 55, 87
  sundry items, 120–132
  tips, 120
  to-date, 40
  total see Total costs
  trade see Trade costing
  tradesmen/labourers, 54
  unit see Unit costing

Cost control, introduction to, 3–6
Cost reconciliation, 112
Craneage, 32, 179
Credits, 121
Cumulative data, 108
Curing concrete, 58
Cycle time, 186, 188, 189

Data bank, 70
Data base systems, 269
Daywork operations, 4, 121, 124, 132
Death benefit, 227
Decision making, 146–147
Demolition, 26
Director's fees, 3
Drainage, 182
Dumpers see Transport on site

Earnings, of contract, 4
  programmed, 9
Earthmoving, bulk, 165, 195
Errors
  in tender, 72
  in workmanship see Making good
Estimating, 6, 68
  analytical, 165
  breakdown of rates, 72
  comparative, 166
  feel of the market, 71
  pricing, 70
  textbook standards, 75
Estimator's outputs, 146
Evaluating options, 142
Excavation, 28, 35, 102
Excess overtime see Non-productive overtime
Expenditure of contract, 4
Expenditure data, 57
Expenditure errors, 108
Extension, 154
External works, 20, 27

Fares, 81, 82, 102, 211, 217, 234
Fatigue allowance, 157
Federation of Swedish Building Employers, 141
Feed-back records, 86, 102–107, 127
Fencing, 27
Final snagging, 21
Financial programme, 9
Financial progress, 9, 86
Fixed price contracts, 227
Flexible paying, 29
Flood damage, 21
Formula, 4
Formwork, 20, 28, 35, 121
Fuel and oils, 66, 86, 148

Gang study, 159
Ganger's allocation sheets, 59, 86
Garages, 31

Gearing, 238
Graduated National Insurance (GNI), 20, 81, 82, 217, 227, 229, 234
Grants
 government, 4
 training, 4
Graphs, 127, 238
 chronocycle, 179
 cycle, 179
Grassing, 27
Greasing time *see* Maintenance time
Gross costing, 66
Gross labour rate, 54

Handover, 21
Head office
 charges, 73
 overheads, 148
Histograms, 27
Historical data, 75
Historical outputs, 68
Holiday stamps, 52, 65, 81, 82, 102, 211, 217, 234
House building, 21, 31
 stages, 124

Incentives, 59, 63, 70, 77, 88, 122, 143, 178, 225, 230, 231, 239
 lack of, 145
 schemes, 230
  advantages of, 230
  disadvantages of, 231
  monitoring, 239
  performance-related, 236
  piecework, 232
  profit-sharing, 231
  time-saved, 234
  type of, 231
 targets, 157
Inclement weather, 35, 36, 81, 108, 216
Inefficiency, site, 145
Insulation, loft, 31
Insurances
 general, 81
 national, 4
Interference by client, 145
Interim payments (valuation), 3, 9, 124
Internal company adjustments to tender, 71

Job card system, 59, 67, 143, 145, 242
Job cost, 134, 135
Jobbing, 6, 133, 237
Joinery
 first fix, 31
 second fix, 31, 33

Kerbs, 27
Keyfold, 272

Labour
 allocation of, 67, 125
 cost, 5, 36, 57
 hours, 35
 preliminaries, 108, 110
 small items of, 120
 standing time for, 126
Labour rate, gross, 54
Labour-to-tradesmen ratio, 54
Lasers, 191
Learning curves, 77
Letting margin, 18
Load factor, 173
Lodging allowance, 65, 81
Losses
 action on, 145–149
 on materials, 13
 reasons for, 145–146

Main drainage, 26
Maintenance time, 100, 102, 123, 216
Making good, 145
Management check-list, 146
Manholes, 26
Manpower requirements, 178
Marginal costing, 132
Market, feel of, 71
Master file, 272
Materials, 13
 handling, 21, 32, 33, 178, 179
 hoisting, 21
 lack of, 145
 losses on, 13
 nominated, 4
 permanent, 5
 scrap, 4
 on site, 129
 temporary, 5
Measured work, 33, 88
Method improvement, 157
Method statement, 35, 40
Method study, 170
Mild steel, 209
Minibus, 81, 217
Miscellaneous payments, 217
Mixer set-up, 71, 179
Monitoring expenditure, 57
Mortar,
 mixing, 20, 123
 supply, 29, 30
Multiple activity chart, 171, 181

National increases, 81, 227
National Working Rules for the Building Industry, 230
Nett costing, 50, 65, 66
Nominated sub-contractors *see* Sub-contractors
Non-productive overtime, 52, 81, 104
Normalising, 154
Numerical code, 61

280 *Index*

Objectives, 147
Offices *see* Temporary buildings
On-costs, 81, 102, 110
Output, 36, 218
 factors affecting, 68
Output data, feedback of, 122, 146
 historical, 68
Output standards, 75
Overheads, 20, 86, 110, 124, 218
 allocated labour, 65, 66, 123
 fixed labour, 19, 80, 104, 122
 fixed plant, 80, 104
 head office, 5, 73
 nett costing, 56
 plant, 64, 65, 66, 126
 site, 4, 5, 21, 36, 57, 65, 148
  fixed, 80
  standards for, 68–79
  variable, 81
 variable, 56, 81, 105–107, 122, 130
  effect of standards on, 130
 wage sheet, 65
Overtime, 211, 218
 allocations, 66, 218, 229

Paths and pavements, 27
Payment for allocation, 63
Performance, 101, 236, 240
Personal needs allowance, 157
Personal observations, 63
Photographic techniques, 179
 memomotion, 180
Piecework, 124, 232
Piling, 28
Planking and strutting, 121
Planning
 lack of, 145
 meetings, 143, 244, 246
 short-term, 145, 244
 weekly, 246
Plant
 allocation of, 125
 charges, 4, 5, 57, 63
 costs, 37, 86
 hire, 218
 light, 120
 maintenance, 36
 overheads, 64, 65, 66, 126
 preliminaries, 110
 small items of, 120
 standing time for, 145
 written off, 71
Plus rates, 82
Policy allowance, 178, 236, 239
Porches, 31
Post costing, 143
Pre-cast units, 29
Pre-costing, 144, 250
Predetermined motion time systems,
 (PMTS), 166
Preliminaries, 4, 71, 80, 108

Premium time *see* Non-productive overtime
Pricing
 estimator's, 68, 70
 textbook, 68
Prime costs, 3, 7–11
 breakdown, 9
 calculation, 11
 summary, 8
Process charts, 178
Production control, 77
Productive work, 20, 21, 109
Profit and loss account, 3
Profit margin, 73
Profit sharing scheme, 231
Profitability, 3
Program generators, 269
Programmed earnings, 9
Public holiday, 81, 82, 102, 108, 216
Pumps, 64, 72

Questioning technique, 170, 173
 primary questions, 171
 secondary questions, 171
Quotation, 133

Random activity sampling, 167
Rate refunds, 4
Rating
 blanket, 182
 clinics, 153
 observed, 153, 185
 standard, 152
Record card, workman's, 136, 137
Redundancy pay, 81, 82, 217, 220, 227
Reinforcement
 cut and bend, 28
 fixing rod, 28
 high tensile, 209
 mesh, 29
Relaxation allowances, 150, 153, 156, 176, 186
Retention, 8
Rigid paving, 27
Ring beam, 30
Road
 gullies, 26
 on site, 71

Scabbling, 20
Scaffolding, 29, 31, 81
Scrap value, 210
Selective employment tax, 227
Services, 26
Shift work, 211, 225
Sick pay, 81, 82, 217, 227
Site costs, 3, 110
Site inefficiency, 145
Site overheads, 21, 65
 facilities, 21
 fixed, 80
 standards for, 68–79
 variable, 81